Electronic Controls of Diesel Engines

SP-673

The papers included in this volume are abstracted and indexed in the SAE Global Mobility Database

NEW ENGLAND INSTITUTE
OF TECHNOLOGY
LEARNING RESOURCES CENTER

Published by:
Society of Automotive Engineers, Inc.
400 Commonwealth Drive
Warrendale, PA 15096
September 1986

Permission to photocopy for internal or personal use, or the internal or personal use of specific clients, is granted by SAE for libraries and other users registered with the Copyright Clearance Center (CCC), provided that the base fee of $3.00 per copy is paid directly to CCC, 21 Congress St., Salem, MA 01970. Special requests should be addressed to the SAE Publications Division. 0-89883-944-0/86 $3.00

ISBN 0-89883-944-0
SAE/SP-86/673
Library of Congress Catalog Card Number: 86-082258
Copyright 1986 Society of Automotive Engineers, Inc.

PREFACE

The use of electronically based devices for the control of fuel delivery and ignition has become dominant in the passenger car gasoline engine. The same forces which brought about the changes in the car business are now being felt by the engineers designing and developing vehicle diesel engines; i.e., the legislative pressures for reductions in exhaust emissions and noise and the need to maximize the efficient use of fuel oil.

The increase in development and research activity in the area of diesel engine electronic control was recognized by the Diesel Engine Committee of SAE. As a result technical sessions devoted to the special needs of diesel engine systems were organized for the 1985 and 1986 Detroit Congress. The high level of interest stimulated by these sessions indicated the need for a lasting reference document and the eleven papers prepared for the foregoing meetings are now assembled in this volume.

The papers contained herein provide a well balanced view of electronic controls representing both the equipment and engine manufacturers' experience and covering design and application of this technology to diesel engines. Additional depth is given by the diversity of sources which includes work done in the U.S.A., Japan, and Europe.

R. E. Turney
Ford Tractor Operations

TABLE OF CONTENTS

850169   **A New Electronically Controlled Injection Pump for Diesels** 1
         A. Maynard
            Stanadyne Diesel Systems
               Hartford, CT

850170   **High-Pressure Injection Pumps with Electronic Control for Heavy-Duty Diesel Engines** 15
         R. Schwartz
            Robert Bosch GmbH

850172   **Development of a Fully Capable Electronic Control System for Diesel Engines** 29
         Makoto Shiozaki, Nobuhito Hobo, and Ichiro Akahori
            Research & Development Dept.
               Nippondenso Co., Ltd.
                   Kariya City
                      Aichi-pref. Japan

850173   **Caterpillar 3406 PEEC (Programmable Electronic Engine Control)** 37
         Michael E. Moncelle and G. Clark Fortune
            Caterpillar Tractor Co.

850174   **A Digital Control Algorithm for Diesel Engine Governing** 53
         Daniel C. Garvey
            Woodward Governor Co.
               Engine and Turbine Controls Div.
                   Fort Collins, CO

860141   **A New Concept for Electronic Diesel Engine Control** 63
         Alois Kainz, Artur Seibt, and Christian Augesky
            Voest-Alpine Friedmann GmbH
               Vienna, Austria

860142   **Exhaust Emissions Influenced by Electronic Diesel Control** 73
         E. Gaschler
            Res. Div.
               Volkswagen AG
                   W. Germany

860143   **Digital Self-Calibrating Hall-Effect Sensor for Electronically Controlled Engines** 81
         Michael S. Ziemacki and George D. Wolff
            Wolff Controls Corp.

860144   **Electronic Control of Diesel In-Line Injection Pump — Analysis and Design** 91
         Kazuro Nishizawa, Hiroshi Ishiwata, and Kenji Okamoto
            Diesel Kiki Co., Ltd.

860145   **The Second Generation of Electronic Diesel Fuel Injection Systems — Investigation with a Rotary Pump** 107
         Keiichi Yamada and Hidekazu Oshizawa
            Diesel Kiki Co., Ltd.

860146   **The Electronic Governing of Diesel Engines for the Agricultural Industry** 121
         Peter Howes, David Law, and Dalip Dissanayake
            Lucas CAV Ltd.

850169

A New Electronically Controlled Injection Pump for Diesels

A. Maynard
Stanadyne Diesel Systems
Hartford, CT

ABSTRACT

An electronically controlled rotary-type diesel fuel injection pump has been developed to provide full authority fuel and timing control for light-duty diesel engines in automotive, agricultural and industrial engine applications.

Vehicle testing to date on a 4.3 liter V-6 and a 6.2 liter V-8 diesel engine has resulted in obtaining superior performance in several areas when compared with engines equipped with their original mechanical fuel injection pumps.

An eight-bit, single chip microcomputer, and linear stepper motors, along with engine and vehicle mounted sensors provide the full authority electronic control functions.

INTRODUCTION

For the past five years, Stanadyne's Diesel Systems Division has initiated several programs for evaluating the application of electronic controls to rotary distributor pumps. These programs range from limited authority control primarily used for trim adjustments to complete "fly by wire" full authority control. In the full authority category, two types of pumps are being investigated. The first employs a solenoid valve to control the amount of spill and the instant at which it takes place to achieve fueling and timing capability. The second is the Plunger Control Full Authority (PCF) pump which is the subject of this paper.

A major goal of this program is to retain as much of the DB type design as possible and limit changes to those which are necessary for implementing an electronic control mode of operation. The pump resulting from this criterion has only one major change which is found in the metering section. Inlet metering, normally used for fuel control in these pumps, has been replaced by a stroke control mechanism.

The PCF pump is designed for use on current DB2 applications including automotive, agricultural and industrial engines. As the label "full authority" implies, all scheduling of fuel and timing is done through a microcomputer using sensor information to control actuators on the pump. No mechanical connection exists between the operator and the pump.

Vehicle testing is being conducted to accumulate reliability and performance data on the pump, microcomputer, sensors, wiring harness, connectors and alternate control strategies. An automobile with a 4.3L V-6 engine and a light-duty truck with a 6.2L V-8 are being used for the test vehicles.

The demand for electronic control of pump functions is growing to meet the increased demands for functional improvements in engines. This is demonstrated by one approach of adding discrete electrical devices to handle specific pump functions over a limited range.[1] As these devices expand towards complete control over a function and therefore to a more complex mode of operation, a crossover point will be reached where full authority electronic control represents the best integrated solution for satisfying increasingly stringent requirements.

PROGRAM OBJECTIVES

Pumps which employ electronics for control functions have a superior level of flexibility when compared with mechanical pumps. Since flexibility is a key for future pump applications, this factor becomes especially important in the early development of a system for electronic pump control. The greater the flexibility at the beginning of a program, the better the chance for the pumps to extend their usefulness over many applications. Flexibility can be assured by making the system a software-oriented application which is easily adapted to a wide range of engines. Substantial savings may be realized in development time and parts inventory if an emphasis is placed on modifications of software rather than hardware.

A list of general program objectives upon which the PCF pump was designed is given below.

1. Small size, light weight, low cost, configured to be mounted in current and foreseeable future applications.
2. Have the capability for evaluating many types of control strategies and systems.
3. Have the flexibility to utilize a wide range of actuators and sensors.

The specific objectives for this pump are shown below. Those which have been tested to date in vehicles are indicated by an asterisk (*).

A. Fuel delivery control functions
   *1. Maximum and part load fuel shaping
   *2. Excess fuel
   *3. Throttle progression tailoring
   *4. Min.-max. or all speed governing
   *5. Cold engine warmup scheduling
   6. Altitude fuel compensation
   7. Turbocharge boost compensation
   8. Transient fuel trimming
   9. Feedback signal for more complete emissions control scheduling

B. Timing control functions
   *1. Speed-load advance
   *2. Cold start advance scheduling
   *3. Timing map flexibility
   *4. Pump installation error reduction
   5. Drive shaft wear
   6. Altitude timing compensation
   7. Emissions level control

C. Additional vehicle functions
   *1. Cruise control
   *2. Transmission lockup release points (torque converter clutch-type transmissions)
   *3. Exhaust gas recirculation control
   4. Glow plug control
   5. Optimal control of total drive train package
   6. Diagnostics of all electronically operated components
   7. Information for dashboard displays

The objectives selected for implementation were those which are adaptable to the vehicles available for this program. Since these are production vehicles, they lack the sophisticated components necessary to increase the number of objectives realized.

PUMP DESCRIPTION

The control of fuel delivery is accomplished by a stroke control mechanism rather than a metering valve traditionally found in DB type pumps. It was necessary to replace inlet metering because of the large amount of speed bias affecting delivery as shown in figure 1A. An additional characteristic which must be considered is the nonlinearity of delivery with position for constant speed especially in the low delivery range. A control system using a pump with these delivery characteristics can introduce problems related to the increased range of dynamic loop gain needed to insure accuracy and stability throughout the engine's operating enve-

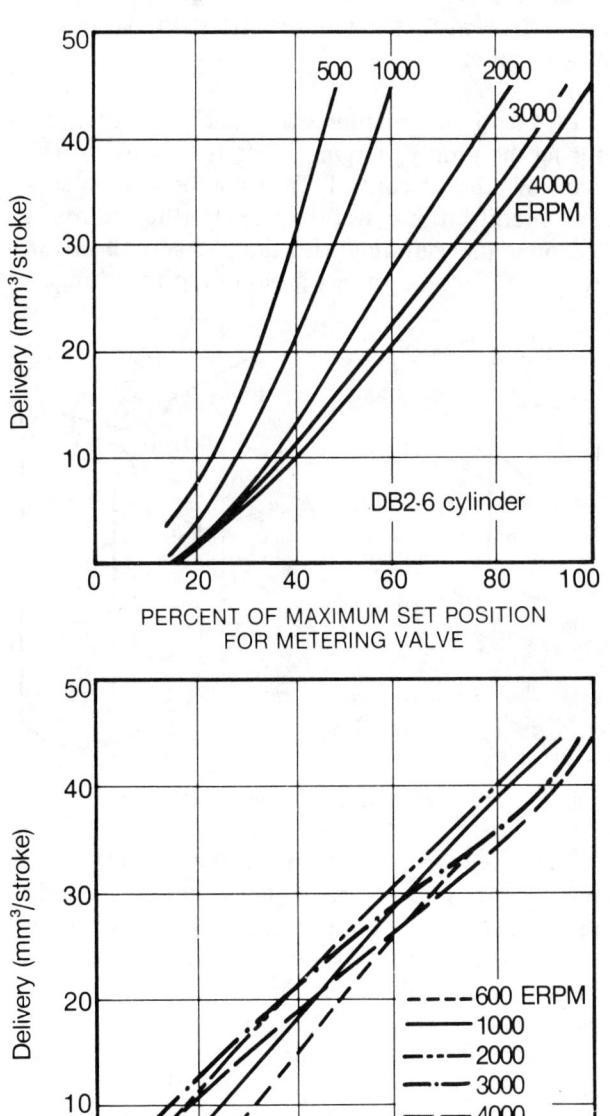

Fig. 1A & B — Delivery characteristics of inlet metered and plunger stroke control pumps.

lope. These problems would have a tendency to be magnified in a production build. The plunger control fuel delivery characteristics shown in figure 1B are more suited for electronic control due to the smaller speed bias and better linearity between the controlled position parameter and delivery. This improved pumping characteristic is a result of the charging sequence taking place closer to the transfer pump pressure (no pressure drop across a metering valve) and the plungers being limited in their outward radial displacement. Under these conditions, a precisely controlled quantity of fuel is trapped between the plungers prior to the beginning of pumping.

Control of the plungers' radial displacement during the pump charging phase is the basic operation of plunger control. To accomplish this, several modifications were made to the pumping components beginning with the plungers shown in figure 2A. Two ramps are ground into the plungers

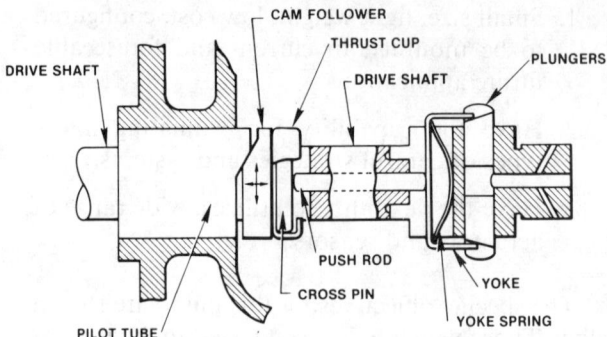

Fig. 3 — **Components for axial positioning of yoke.**

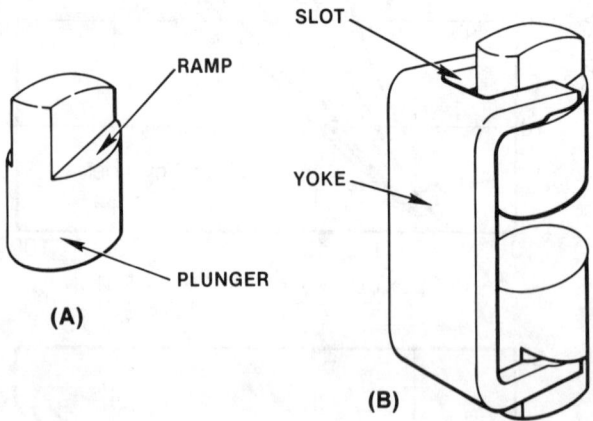

Fig. 2A & B — **Plungers and yoke assembly.**

each having the same inclined plane with respect to the centerline axis. The angle of the ramps is dependent on the amount of radial plunger displacement needed to get the required delivery and the incremental resolution in the case of a device such as a stepper motor. As shown in figure 2B, these ramps engage with the fingers of a yoke which straddles both plungers when assembled in a rotor. The slots between the yoke fingers serves to keep the plungers from rotating and permits them to be pushed inwardly during the pumping sequence.

The yoke and plungers assembled in the rotor as shown in figure 3 permits the axially positioning of the yoke which controls the radial displacement of the plungers during charging. A slot in the driven end of the rotor locates and guides the yoke's axial movement. From figure 3, moving the yoke to the left will increase the amount of fuel trapped between the plungers during charging which results in increased delivery. A yoke slot machined in the driven face of the rotor permits free radial yoke movement to automatically shift to accommodate uneven outward and inward displacement of paired plungers. Since pumping begins only when the plungers move inwardly simultaneously, the yokes self-centering or radial freedom of motion prevents an outward force on the yoke from uneven inward plunger actuation. An additional benefit due to this self-aligning feature is the initial rate of injection is compelled to remain constant for a given delivery.

The positioning of the yoke, and therefore the control of fuel delivery, is accomplished in the following manner. Referring to figure 4, a profile ground into the pilot tube engages with a second profile ground into the cam follower. Rotational movement of the cam follower causes the ramps to move relative to each other resulting in an axial displacement between the pilot tube and the cam follower. This motion is transmitted by the thrust cup, cross pin and push rod to the yoke. Both the thrust cup and the cross pin rotate with the drive shaft. The cam follower only rotates when a change in the fuel delivery is made. A yoke spring serves

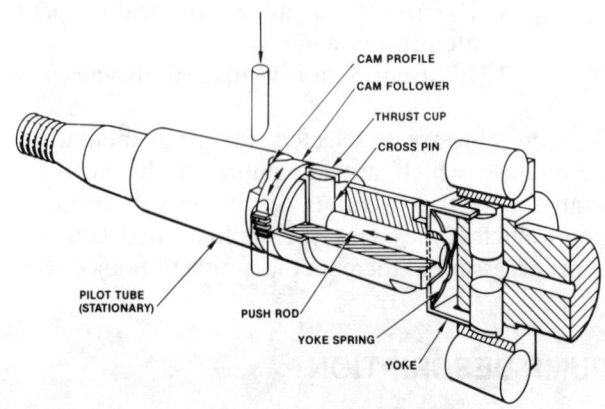

Fig. 4 — **Cam follower rotation controls yoke position.**

to insure that the yoke remains against the push rod and helps to limit any rocking movement of the yoke.

When pumping occurs, the rollers, shoes and plungers are forced inwardly by the lobes on the pumping cam. The plungers' movement causes a disengagement between the plunger and yoke ramps. To prevent any contact between the shoes and yoke, the maximum distance across the outside surface of the yoke's fingers must be less than the minimum distance between the shoes at the "over the nose" cam position.

During the initial design phase, it was thought that the pressure loading forces on the plungers would be light enough to permit a stepper motor to directly position the cam follower as shown in figure 5. The stepper motor would simply position the fuel rack to rotate the cam follower. This motion would be transmitted by a feedback rack to a position sensor, currently a potentiometer, for a fuel level signal. In the initial phase of testing, the plunger forces during the charging event were found to be greater than anticipated. A pulsating force is created which acts on the yoke and in turn is transmitted by the push rod to the cam follower. The result of this action is that the fuel level signal contains a superimposed dither signal on the actual fueling level signal. Two major problems arise because of this dither signal. First, the amount of corrective action by a control system is increased and second, there is a greater potential for excessive wear of components comprising this section.

This pulsation problem was further compounded when testing showed the magnitude of the force created was in the range of the maximum force generated by the stepper motor. With this configuration, one could expect at best a slow system response for fuel control due to the inability of the stepper motor to follow the rapid step pulse commands. Increasing the rotor diameter of the stepper motor to create more force output would not be a satisfactory solution. The additional rotor inertia would decrease the dynamic response of the stepper motor and require a lower step rate to prevent missed steps.

This problem was solved by modifying the fuel control section to incorporate a servo system as shown in figure 6. Component sizing was based on

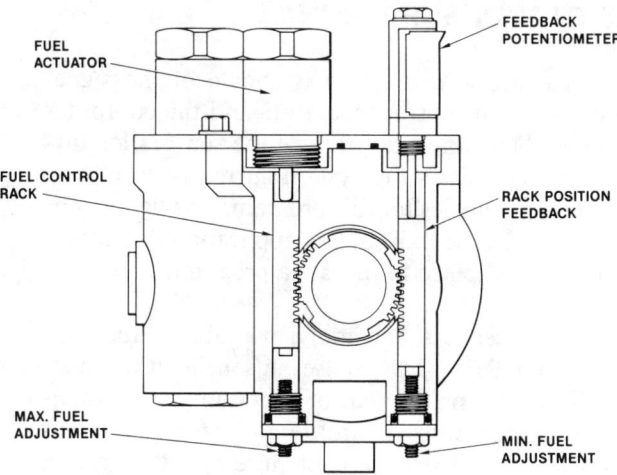

Fig. 5 — Stepper motor acting on fuel control rack.

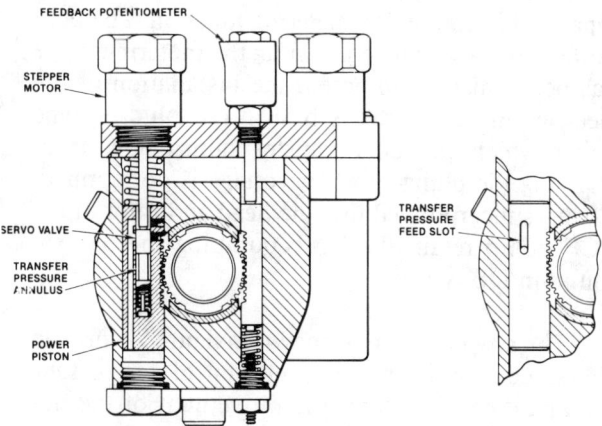

Fig. 6 — Stepper motor acting on servo valve to control fuel delivery.

providing the necessary spring force to the feedback rack which eliminated the pulsations while giving the speed response necessary for idle and full throttle transients. Vehicle testing verified that the response of the servo system is adequate. A difference in engine transient performance cannot be detected when compared to the original mechanical pump.

The servo system can be characterized as a force amplifier where position of the power piston follows the servo valve position very closely. Since the servo valve is completely balanced, a very small force is needed to position it. In this case, a small spring at the base of the servo valve lightly forces it against the shaft of the stepper motor. Keeping this force very small allows the stepper motor to operate under conditions more favorable for high speed response and durability.

Positioning of the servo system takes place in the following manner. Referring again to figure 6, movement of the servo valve upward allows fluid at transfer pump pressure to flow into a passage leading to the bottom of the power piston. As pressure increases in this chamber, the power piston moves upward until its porting lines up with the lands of the servo valve. At this point, the pressure and load forces on the piston reach an equilibrium and upward motion ceases. Downward movement of the servo valve relieves the pressure at the bottom of the power piston by allowing the fluid to exhaust to housing pressure. Springs acting on the top of the power piston and at the base of the feedback rack force the power piston downward until its porting lines up with the lands on the servo valve.

In the PCF pump, fuel cut off is completely separated from the fuel metering function. The bore in the head normally containing the metering valve has been enlarged to accept the installation of the electric shut off. This is basically a plunger type valve which prevents transfer pump flow from charging the plungers when voltage is not applied. Changes incorporated into the head and rotor of the PCF pump retain the basic pumping mechanism found in DB type pumps.

The advance system shown in figure 7 works in the same manner as the fuel control servo. One major difference is that the movement of the advance piston in the retard direction is done by the force of the rollers hitting the cam lobes. Again, a stepper motor is used to position a servo valve which in turn positions the advance piston.

A model of the PCF pump in its eight cylinder configuration is shown in figure 8. The stepper motors for fuel and timing control, the feedback sensor for fueling level and the electric shut off are

Fig. 8 — Eight cylinder PCF pump.

located on the top of the pump for easy access. This configuration permits the pump to be mounted where DB pumps are presently installed.

SYSTEM DESCRIPTION

In figure 9, a diagram is shown of the signal flows between sensors, actuators and the computer module. With the exception of the sensors for turbocharging, this is the system currently employed on the test vehicles for this program. The figure does not represent the limit of this application, but rather a convenient place to initiate a program.

Input sensors include a variable reluctance pickup for RPM/TDC pulses, a sensor in the nozzle for indicating the start of injection (SOI), potentiometers for desired throttle and actual fuel load signals, a thermistor for coolant temperature and a vacuum sensor for feedback of the EGR control. A second speed signal from the vehicle drive shaft is used for the cruise control mode of operation. Manifold pressure and air temperature are used in turbocharge applications. On/off signals are sent to the computer by the ignition switch and the cruise control switches.

Output signals from the ECU consist of a pulse train to the two stepper motors for fuel and timing control, a positive voltage level to hold the ESO open and signals to the solenoids which gate the vacuum for controlling the EGR valve. The ECU also controls when the transmission's convertor clutch disengagement takes place.

TDC and RPM are derived from a single magnetic pickup sensing equally spaced teeth on a tim-

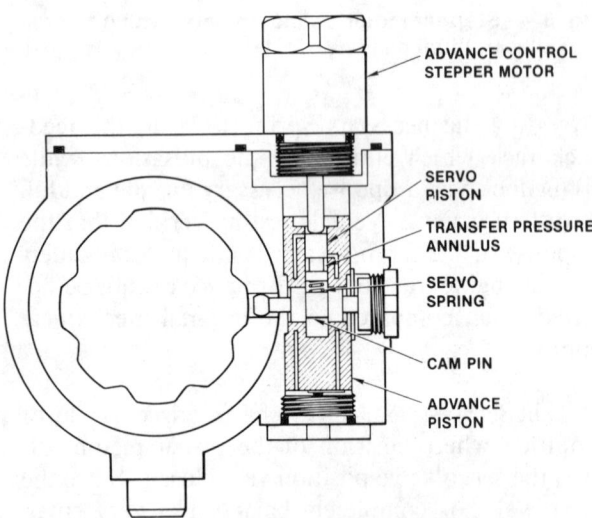

Fig. 7 — Stepper motor controlling cam advance.

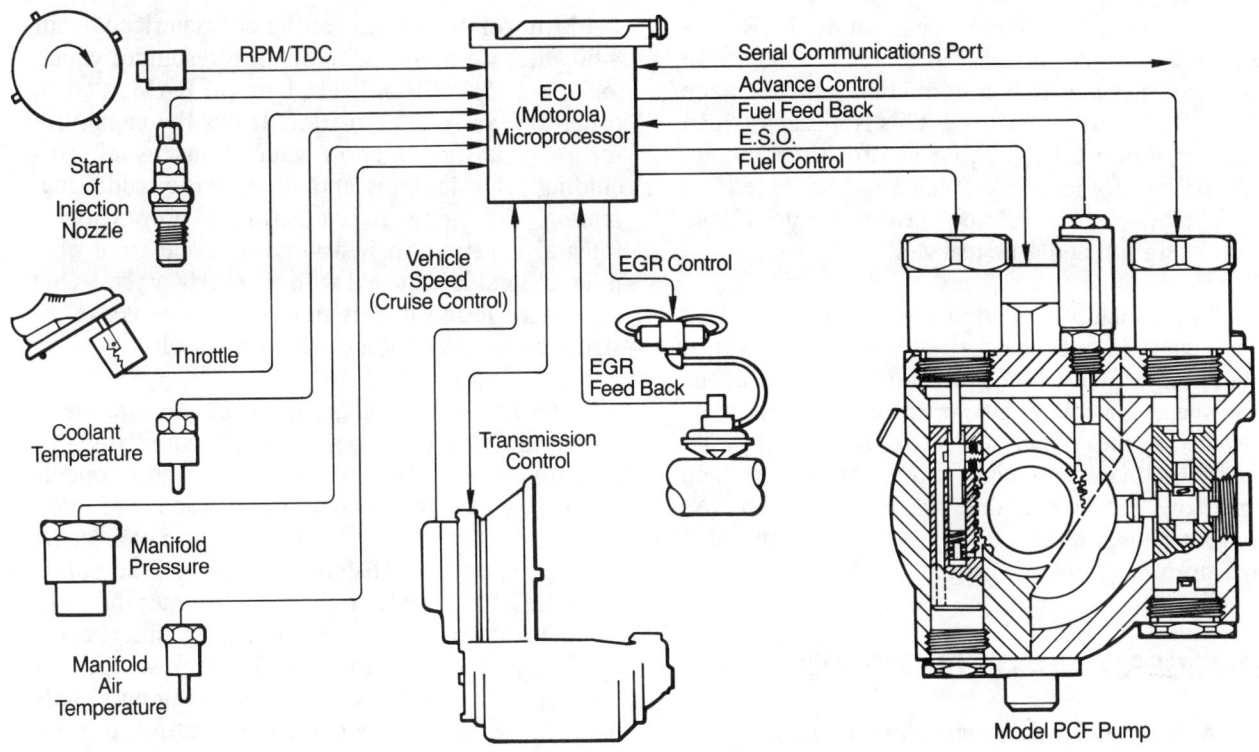

Fig. 9 — Signal flow diagram for PCF pump.

ing wheel. For six and eight cylinder engines, the number of teeth is three and four respectively. The gear is mounted to the crankshaft and is angularly adjusted to obtain a specific phase difference between the mechanical TDC and the sensed pulse from a tooth. The magnitude of this phase difference is equal to one half the angle between adjacent teeth. For engines with six and eight cylinders, this angle is 60 degrees and 45 degrees respectively. Employing this phase difference insures that the sensed pulse from which TDC is calculated and the start of injection signal occur far apart from each other. This helps in maintaining accurate engine timing and RPM calculations at high engine speeds. The tooth offset also provides a convenient method for determining SOI when it takes place in advance of the mechanical TDC occurrence. Pulses formed by the wheel teeth passing under the magnetic pickup are transmitted to the microprocessor which measures the time between pulses for calculating RPM. The time difference between SOI and TDC both having the previous pulse as a reference is used in conjunction with the time difference between RPM pulses to determine the advance angle. This advance angle is the feedback information which is compared to a desired advance to form an error upon which a correction can be made. To date, this method has given very satisfactory results. However, there are probably engines which require more resolution in RPM data. In this case, a separate RPM wheel containing more teeth in addition to a TDC wheel may be needed.

Start of injection is measured by a Hall effect transducer installed in the nozzle to detect the moment of valve opening. An alternative sensor currently being evaluated makes use of piezoelectric ceramics mounted in the nozzle body to read the instant of valve opening. Timing adjustments are made based on the difference between SOI and TDC and the desired timing map stored in the computer memory. While this represents the controlling of an engine state somewhat removed from the combustion process when compared to an optical pickup, control timing schedules can be defined which provide very satisfactory performance. Fuel delivery is measured indirectly by a potentiometer sensing the position of the fuel feedback rack. As was pointed out previously, stroke control gives a good linear relationship between rack position and delivery. Consequently, as a result of this correlation, the necessary condition exists for good control accuracy.

EGR is controlled through a dual solenoid valve which modulates the vacuum applied to the diaphragm of the EGR valve. The relation between the amount of EGR and vacuum level is stored in the ECU for use in the feedback calculation of the EGR control routine. A vacuum sensor in the EGR vacuum line provides a signal to the ECU to generate the error upon which a correction can be made through the solenoid valves. This is basically open loop operation for emissions control since the actual state is not fed back. If the appropriate sensors were available for tailpipe measurements, closed loop operation could be realized.

The coolant temperature serves to provide a bias signal on both the desired idle speed and timing of the engine. A schedule in the ECU raises the idle speed for cold engine operation and gradually lowers it as the coolant warms up. Timing for a cold engine is biased to have more advance to help emissions and give a smoother running engine. As the engine warms up, timing is gradually retarded to a normal advance schedule.

MICROPROCESSOR DESCRIPTION

A M6805R3 microcomputer has been used to date for the controller in this system. R. Martinsons[2] has given a complete description of this microprocessor as it is applied to automotive diesels. Some of the highlights of the M6805R3 are given below.

Motorola's 6805R3 can be characterized as an 8-bit microcomputer with on-chip resources which include 3.75K of ROM, 112 bytes of RAM, a clock timer and an A/D convertor. It has the capability for I/O transfer function calculations usually including table lookups and interpolation, filtering, scaling, averaging and a form of proportional/integral control. An instructional cycle time of 1 microsecond is achieved with an 8MHz crystal. One advantage found in this microcomputer is the instructions available for bit manipulation.

The hardware configuration for the module is shown in figure 10. A power supply conforming to automotive specifications permits system operation below 6 volts. Transient protection is provided to handle short circuits, load dumps and overloads. All high speed digital signals which include pulses from the RPM/TDC and SOI sensors enter the CPU through interrupt pins. The analog channels carry the feedback signals from the fuel rack sensor and the EGR vacuum sensor. Slowly changing signals from the sensors for coolant temperature, throttle and in the case of turbocharging, manifold pressure and temperature are externally multiplexed for transmission through another analog channel to the microcomputer. Digital inputs of very low speed such as those from the ignition switch and the cruise

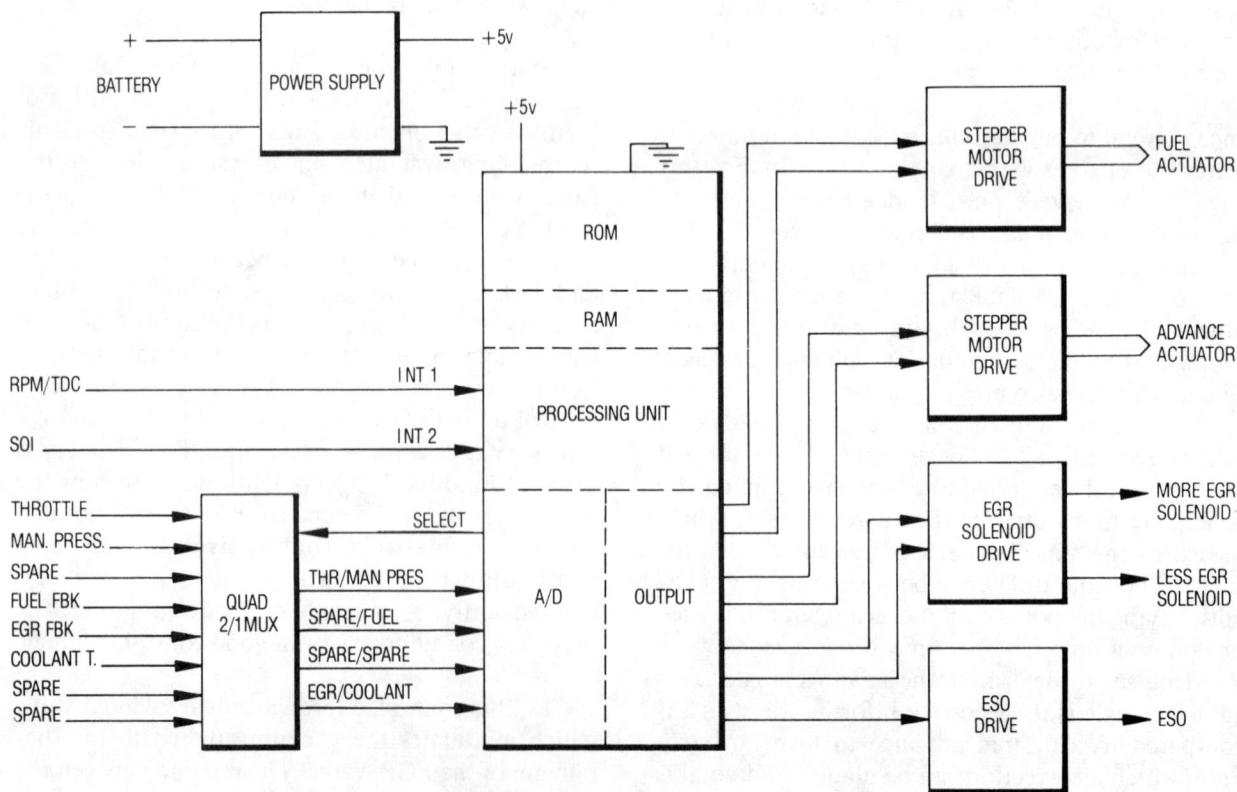

Fig. 10 — Hardware configuration schematic for PCF pump.

control switches are connected to the I/O ports of the microcomputer.

Outputs from the microcomputer include gated signals to the stepper motors and on/off signals to the ESO, EGR control solenoid valves and transmission lockup relay. The drivers for the stepper motors contain the logic circuitry and transistors for gating current to the windings in a properly phased manner to achieve the desired direction of motion.

Simple on/off signals needed for the remaining outputs permit the use of logic/predriver devices which insure the proper output states in the event of a processor failure. The predrivers are followed by power transistors which control the system's actuators.

Software is divided into foreground and background sections. The interrupt service routines are contained in the foreground and are primarily dedicated to timekeeping and input/output operations. Calculations for the actual controlling are done in the background section. Having an architecture of this form prevents interference between the acquisition and control tasks taking place in background and the simultaneous interrupt driven processing taking place in foreground.

The requirement of an operating temperature range from −40 to 85 degrees C for passenger compartment mounting is achieved by this module. Power dissipation is approximately 1.5 watts through the circuitry which results in a temperature rise of approximately 5 degrees C within the enclosure. Pump mounting of the electronics is a distinct possibility for the future.

Fig. 11 — Vehicle installation of emulator display unit.

The exterior components are connected to the microcomputer through a thirty-five pin connector. Careful application of grounding techniques and design of the enclosure prevents electromagnetic radiation from other sources from interfering with the microprocessor.

Finally, the microcomputer can be fully integrated with an emulator system designed for automotive installation. With the increased flexibility provided by the emulator system, development work proceeds at a faster pace and in a more thorough manner. Figure 11 shows the emulator display unit as mounted in the vehicle which allows observation of vehicle and computer data while the vehicle is undergoing road testing.

CONTROL SOFTWARE DESCRIPTION

An overall view of the software strategy implemented in the PCF control system is shown in figure 12. Since the engines of the vehicles selected for this program were known entities in that their performance characteristics had been previously determined, the initial efforts for software development were directed towards duplicating the vehicle performance obtained with the mechanical pump. Specific routines were written to handle fuel control modes of cold starting, idle, part throttle, maximum delivery limiting, high speed cut off and cruise control. For the advance section, routines were written to handle the timing modes for cold starting, cold and normal running.

The primary goal of the fuel control strategy is to satisfy the driver's demands as transmitted by the throttle sensor. His desires must be integrated with the various modes to provide a smooth transition such as obtained with a mechanical pump. A select process determines the current mode of engine operation through a priority criterion and initiates the call for the specific routine for this mode's control. For instance, when the idle and cruise control routines are inactive, a selection is made between part throttle and maximum fuel limiting based on which has the lesser value. The values for part throttle and maximum fuel are determined from lookup tables. The most sensitive transition takes place going from idle to part throttle. Judicious tailoring of governor and part throttle curves at this intersection attenuates the engine's response to this discontinuity.

In the advance strategy, the primary goal is to maximize engine performance (idle quality, tran-

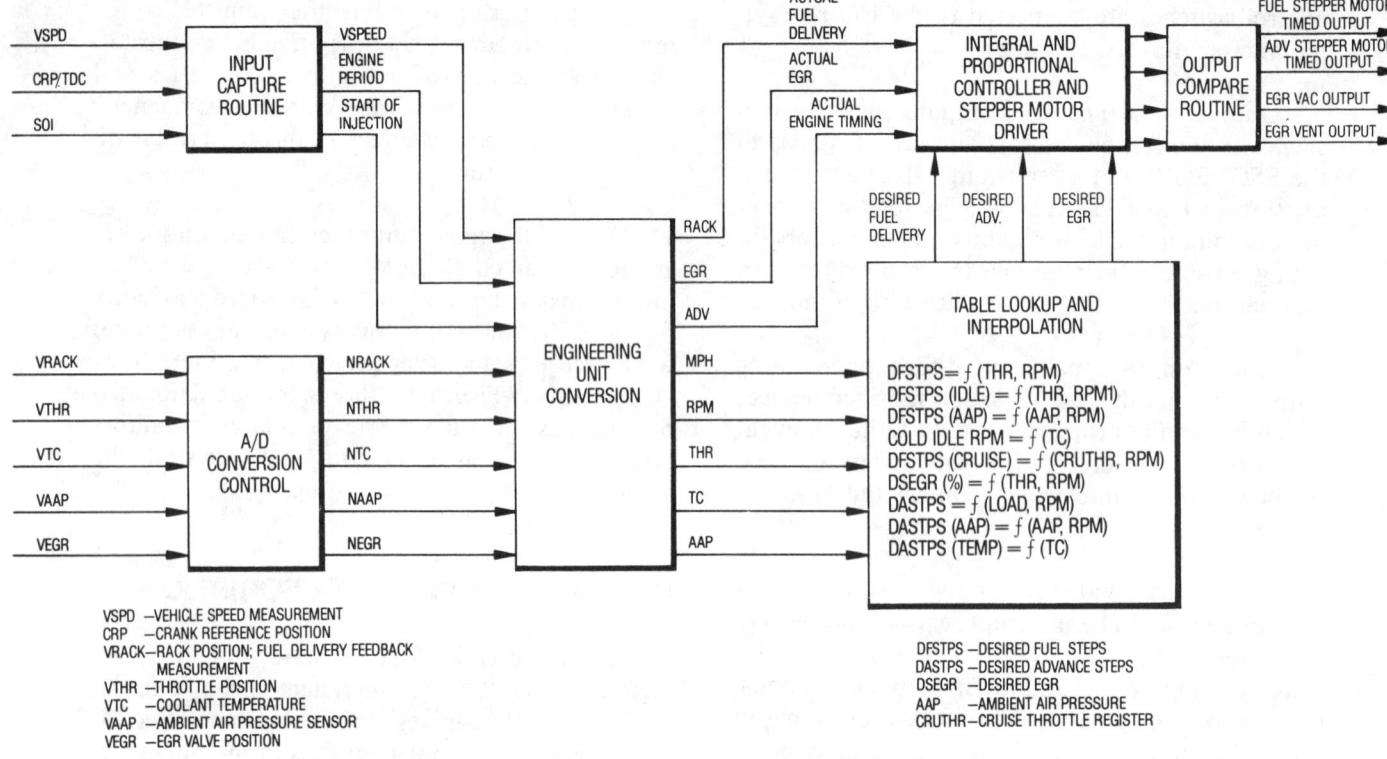

Fig. 12 — Software strategy diagram for PCF pump.

sient response, low noise level) and minimize emissions (HC, NO_x, particulates). Emissions reduction must be done in conjunction with the EGR control. The advance is scheduled as a function of RPM and fueling level with a bias from the coolant temperature. Feedback from the nozzle gives an SOI signal which is compared to the TDC signal to form the correction error.

The EGR strategy is different for the two vehicles used in this program. For the 4.3L application, EGR valve position is derived from a table relating vacuum level to throttle angle and engine speed. In the 6.2L application, only throttle angle is used to switch the EGR valve on and off. Consequently, no feedback is used in this system. Both strategies reflect the control methods used on the mechanical pumps normally found on these vehicles.

A limited failsafe strategy has been implemented in the software chiefly to handle failures which could cause engine runaway or those which can be modified to form a limp home mode of operation. Lack of a pulse from the RPM/TDC sensor will disable the ESO. A possibility exists of using the SOI signal for RPM determination. However, this would require the advance to be held in a nominal position and a separate routine to be used for a deceleration from high speed where the governor would cut off fuel. When the throttle sensor goes out of limit, the program defaults to a 20 percent maximum throttle input.

The loss of the SOI signal causes the advance to remain stationary in its last position. Table I summarizes the failsafe program which would be incorporated in the system shown in figure 9.

VEHICLE PUMP INSTALLATION

Figure 13 shows the PCF pump installed on the 4.3L V-6 engine. The pump takes up less volume

Fig. 13 — PCF pump installed in 4.3L vehicle.

SENSORS		
FAILURE	FAILURE DETECTION	ACTION
Speed	Signal lacking Out of range Initialization check	Disable ESO
Fuel rack potentiometer	Signal lacking Out of range Initialization check	Disable ESO
Start of injection	Signal lacking Out of range Initialization check	Set desired advance Limit fuel delivery Limit speed
Throttle	Signal lacking Out of range Initialization check	Set higher idle Default to lower max. throttle setting
Coolant temp.	Signal lacking Out of range Initialization check	Set signal to normal temp. level
EGR vacuum	Signal lacking Out of range Initialization check	Set for EGR closed
Manifold temp.	Signal lacking Out of range Initialization check	Set for low normal temp.
Manifold press.	Signal lacking Out of range Initialization check	Set max. delivery to altitude limit
Transmission speed signal	Signal lacking Out of range Initialization check	Disable cruise control
ACTUATORS		
FAILURE	FAILURE DETECTION	ACTION
Fuel rack unresponsive	No response to change of desired input	Disable ESO
Fuel rack erratic	Range check	Lower max. fuel level
Advance unresponsive	No response to change of desired input	Lower max. fuel level
Exhaust gas recirculation	No response to change of desired input	Lower max. fuel level
CPU		
FAILURE	FAILURE DETECTION	ACTION
Software	Internal check Watchdog timer	Disable ESO
ENGINE		
FAILURE	FAILURE DETECTION	ACTION
Overspeed	Exceeds high speed limit	Disable ESO

Table I

between the V and is approximately 1.5 inches shorter than the mechanical pump. More room exists under the crossover manifold due to the reduced height of the pump. Figure 14 shows a size comparison between the PCF and equivalent DB pump. The computer module was installed in the

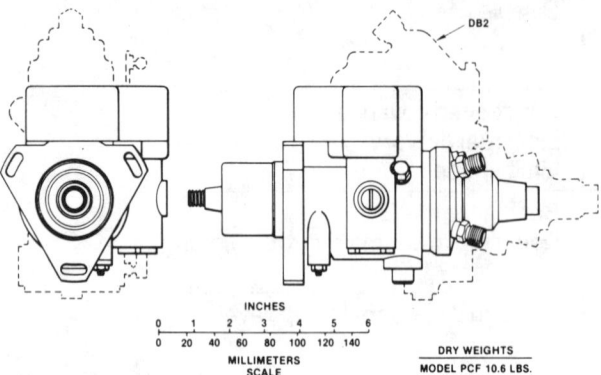

Fig. 14 — Size comparison of PCF and DB2 pump.

trunk to facilitate the ease of accessibility for servicing and making program changes. If accessibility was not a requirement, the computer would easily fit under the front passenger seat. To maintain the vehicle's normal interior appearance, all cables routed through the car are buried under the floor carpets.

The 6.2L V-8 truck installation of the PCF pump is shown in figure 15. For a size comparison,

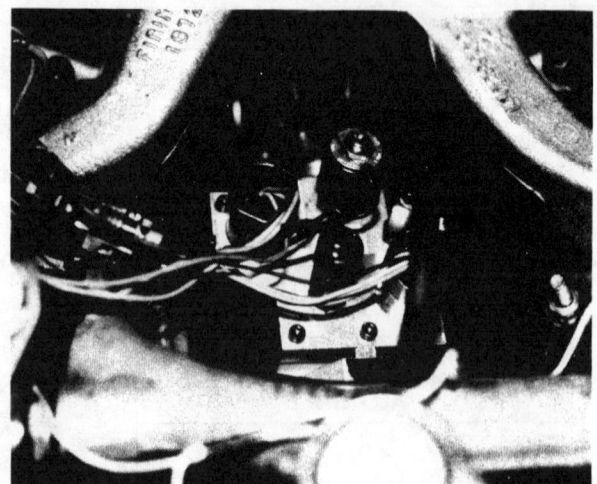

Fig. 15 — PCF pump installed in 6.2L vehicle.

the mechanical pump as mounted on this engine is shown in figure 16. In this application, cabling was constructed to permit the computer module to be installed either under a passenger seat or in the rear of the vehicle's compartment area.

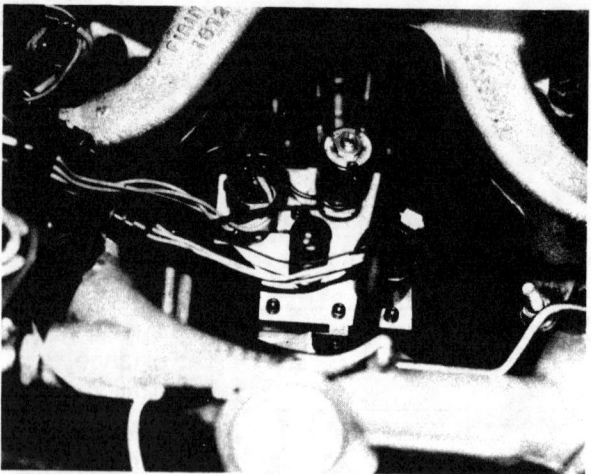

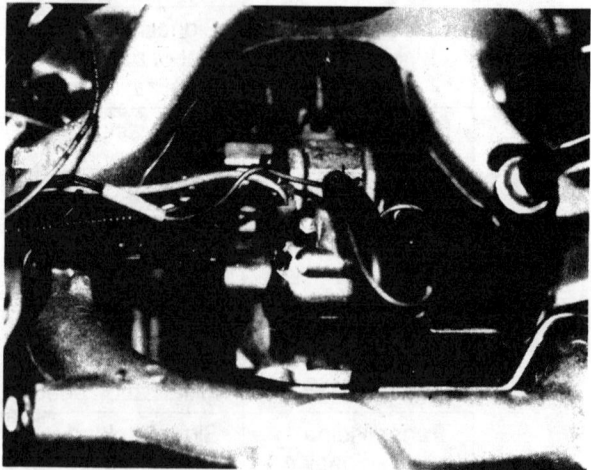

Fig. 16 — Installation of PCF (top picture) and DB2 (bottom picture) in 6.2L vehicle.

VEHICLE PERFORMANCE

Results to date show the vehicle to have very good driveability. One noticeable difference is the smoother idle obtained with the PCF which permits the idle speed to be lowered to 600 RPM. Figure 17 gives a comparison between the idle quality of the original pump and the PCF pump. The use of an integrating servo has been found to provide the responsive actuation necessary for good idle performance as well as transient engine operation. Some emissions testing was conducted on a chassis roll dynomometer which gave encouraging results. However, it became apparent that more extensive emissions mapping of the engine was required to better define the EGR schedule which would optimize the reduction of NO_x while keeping HC and smoke low.

The car has been driven by many people and the comments from them have been favorable.

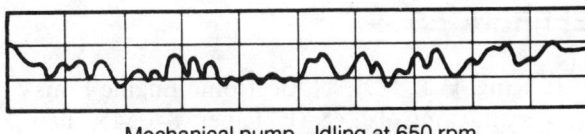
Mechanical pump—Idling at 650 rpm

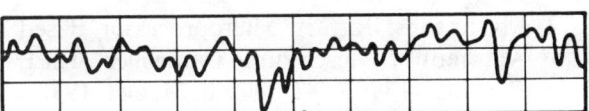
Mechanical pump—Idling at 600 rpm

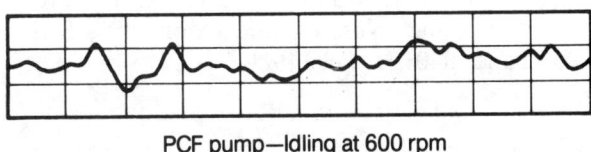
PCF pump—Idling at 600 rpm

Vertical axis—10 rpm/increment
Horizontal axis—.5 sec/increment

Fig. 17 — Idle quality of PCF and DB2 pump.

Most feel that they are unable to tell the difference between the mechanical and the electronically controlled pump.

The car and the truck are used on a regular basis to accumulate mileage. Several long distance trips have been made with the car for demonstrations in Chicago and Detroit. Inspection of the hardware is made periodically and some changes have been made where wear problems exist. Improvements have also been incorporated in the software for evaluation.

CONCLUSION

The day is rapidly approaching where the multiplicity of functions offered by electronic controls will become the practical and economical method for controlling engines operating under stringent conditions. Questions, however, still abound at this time concerning the risks which must be reduced before full authority pumps will become a viable alternative to the highly sophisticated, but compromised, limited authority systems.

For development programs in the early stages of evaluating electronic systems, a pump offering extreme flexibility is desirable to reduce the time required to investigate areas of concern prior to production. The PCF pump is designed to accomplish this goal in several ways. Actuators are not restricted to stepper motors alone. Other types of actuators (PWM solenoids, D.C. motors, etc.) can be installed in place of the stepper motors to evaluate their suitability for use in a control system. Most control functions for engine operation (torque curve shaping, idle governing, timing, etc.) can be changed through software modifications, thereby reducing the test down time. This system has been used to generate information for current and near future programs.

The marketplace will determine the acceptability of electronic fuel systems as applied to diesel engines. How they are perceived in terms of reliability, performance and emissions reduction weighed against cost will govern to the greatest extent the rate at which engine manufacturers apply them to their engines. Emissions regulations will have a significant role in the above process.

An area of great concern having a direct bearing on the rate at which the level of sophistication rises in electronic controls for diesel engines is sensor technology. Without highly reliable and inexpensive sensors, any electronic control system will be constrained from achieving its optimum results. At this time, significant improvement is needed for sensors measuring SOI, fuel rack position and pressure particularly in the areas of unit-to-unit accuracy and endurance.

The use of the microprocessor is expected to expand from a dedicated function of pump control to one of a central controller for handling many engine and vehicle functions. Much of the information now used for control of the pump can be used for driver or operator display data. Solid state dashboard displays are appearing in more applications. They can use raw and calculated information from the microprocessor to display operator information[3]. Redundant sensors can be eliminated by using the computer in this manner and this would lead to a lower total vehicle manufacturing cost to offset the potential for the pump and control system to be more expensive compared to mechanical systems. The possibility exists that life cycle costs of car ownership can be reduced by programming diagnostic routines permitting rapid troubleshooting of powertrain problems. Output from the vehicle's computer used in conjunction with artificial intelligence programming could be used to diagnose engine malfunctions accurately and prescribe the repairs to be made. This represents one approach to an objective of lowering repair costs and time for repairs which owners would certainly appreciate.

ACKNOWLEDGEMENTS

I would like to thank several Stanadyne employees who contributed immensely to the success of this project. Mr. John Pappalardo was the principal engineer in charge of the computer application in this system. Mr. Leighton Keirstead was responsible for the mechanical system and its installation into vehicles. He was ably assisted by Mr. Joe Gura throughout this project. I would also like to thank Motorola's Automotive and Industrial Products Group for their technical consultation help concerning the microprocessor controller.

REFERENCES

1. Ring, H.J., "Diesel Electronic Engine Emission Controls", SAE Paper 840545, Feb. 1984.

2. Martinsons, R., "A Microprocessor Based Automotive Diesel Pump Controller", IEEE Paper CH1799/6/82/0000/0031, Oct. 1982.

3. Long, G. and Korn, H., "An Automotive Electronic Instrument Cluster with a Programmable Non-Volatile Odometer", SAE Paper 840151, Feb. 1984

850170

High-Pressure Injection Pumps with Electronic Control for Heavy-Duty Diesel Engines

R. Schwartz
Robert Bosch GmbH

ABSTRACT

Within the Robert Bosch Diesel Injection Pump Program, the MW and P pumps have a fundamental significance concerning medium-duty and heavy-duty engines.

These engines have developed considerably in the last years with regard to combustion efficiency and emission reduction. In order to reach these new targets, specific developments of the pump were necessary to realize injection pressure of 900 bar and 1200 bar for the MW and P pumps, respectively.

A further development emphasis of the past years was the development of a robust electronic governor concept to take advantage of this new technology. The topics of the paper are:
- Further development of the pump drivetrain and housing to cope with the difficulties inherent in producing 1000 bar.
- Development of constant-pressure valves.
- Preparation of the electronic governor concept for commercial vehicles.
- Development results of the electro-magnetic and electro-hydraulic actuators.

INTRODUCTION

During recent years, the public awareness of the limits imposed upon our energy reserves and the necessity for improved environmental protection has increased considerably.

As a result of higher energy prices and the passing of tough emission legislation in the industrial nations across the world, the activities concerned with the further development of all types of engines in order to improve their environmental compatibility and their fuel consumption have been rigorously intensified.

For HD diesel engines, this results in the following aims in particular:
- further reduction in fuel consumption.
- reduction of black smoke, particularly in the starting and full-load ranges, as well as transient conditions.
- adaptation to the more severe emission legislation with regard to normal emission and particulate emission.
- lowering of the noise level.

The fuel injection equipment must make its own contribution if these aims are to be satisfied. Here, the following main points of emphasis result:
- the use of higher injection pressures in order to improve the fuel atomization while at the same time maintaining the optimum duration of injection.
- the optimization of the fuel injection curve towards achieving a some what flatter ascent and a sharper cut-off. This is particularly impotant with regard to noise reduction.
- provision of an optimum start-of-injection map; if necessary, by applying load and speed dependent injection timing.
- improvement of the adaptation or correction efficiency by the processing of more correction variables.
- further improvement of the fuel metering accuracy.

ON-GOING INJECTION PUMP DEVELOPMENT:

Within the framework of this publication, particular emphasis will be attached to reporting on the on-going development of the model MW and model P Heavy-Duty fuel injection pumps together with the introduction of electronic diesel control (EDC).

The more complex on-going development as applied to the fuel injection nozzles and the timing device has been reserved for a later publication.

The best overview of the current Bosch in-line fuel injection pump program for DI-engines is provided by the rate-of-injection diagram (Figure 1). The maximum attainable injection-pressure-dependent rate of injection is shown for each pump version, taking into account the loading limit imposed by the particular design. The

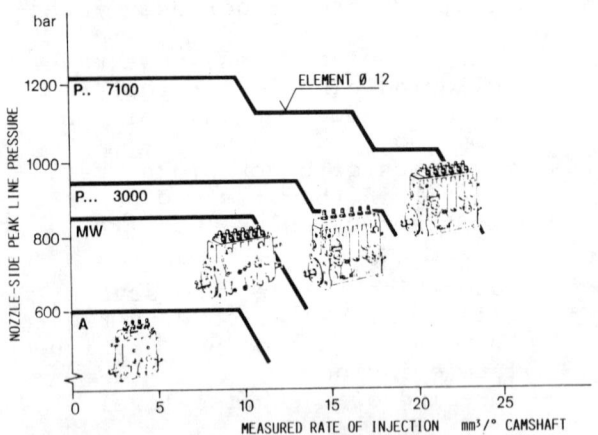

Figure 1 - Hydraulic performance of the A, MW and P fuel injection pumps.

diagram was generated by evaluating a large number of approved applications for direct injection engines. The figures given apply in each case at rated speed.

Compared to the classical A-pump, today's MW pump permits pressures which are about 250 bar higher, and can be used for increased cylinder powers up to about 30 kW.

Until now, the P-pumps were mainly applied for cylinder powers of about 40 kW and peak nozzle pressures of 800 bar. The new P 7100 versions presented here can be used for cylinder powers up to about 50 kW and for nozzle pressures around 1100 bar.

MW-PUMP

Using an MW 6-cylinder pump with mechanical governor as an example, Figures 2 and 3 show the cross and longitudinal sections of the pump.

The characteristics of a compactly designed in-line pump suitable for extreme loading are:
- forged flange-type plunger-and-barrel assemblies.
- low friction control linkage.
- rigid diecast housing, particularly in the area of the driving gear.
- closed tappets.
- stiff, short camshaft.
- non-floating bearing principle.

By applying the finite element method to the optimization of the components, in particular the housing and the flange-type plunger-and-barrel assembly, it was possible to increase the application limit to 850 bar, pump side pressure.

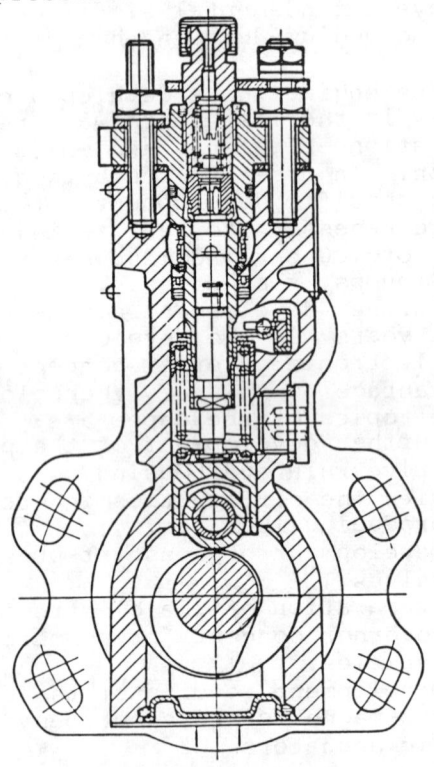

Figure 2 - Cross section of MW 1100 pump.

After completion of thorough testing, the pump proved itself in the field to be reliable and robust. These results were obtained particularly in the USA. The MW pump is gaining in importance for high-performance Medium-Duty engines due to its compact design.

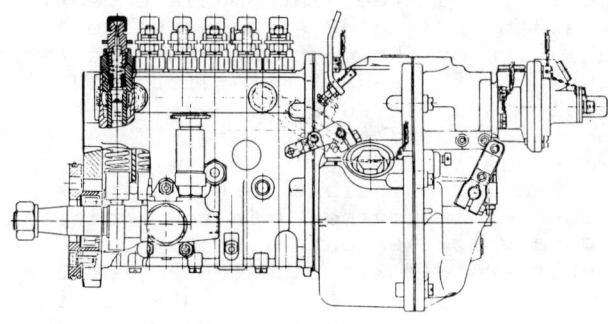

Figure 3 - Longitudinal view of the MW 6-cylinder pump with RQV governor.

P-PUMP FAMILY

Recently the P-pump in particular has been intensively developed towards higher power outputs. Figure 4 shows the trend in the power diagram. The trend towards higher peak pressures and rates of injection can be clearly seen as well as the resulting necessity for larger plunger diameters and cam lifts.

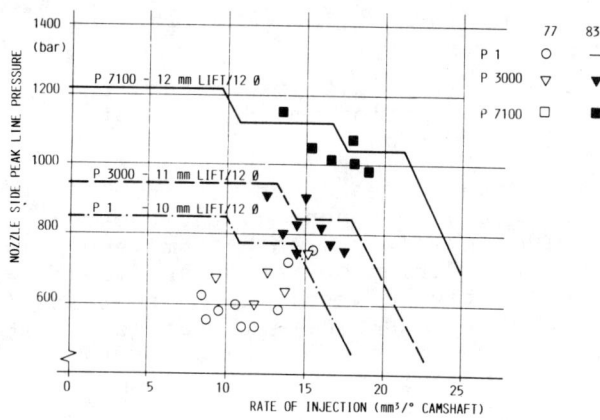

Figure 4 - Application trend of the P-pump family between 1977 and 1983.

Figure 5 shows, by means of a cross-sectional view, a comparison between the P 1/3000 and the latest version of the P 7100, including the main data.

By means of this comparison, it can be seen that if the fuel injection performance is to be increased with regard to quantity and injection pressure, two particular requirements must be taken into account:
- increasing the geometric layout and, at the same time,
- increasing the loading limit of the complete design.

In addition, care must be taken that the efficiency losses due to:
- hydraulic stiffness (dead volume),
- mechanical stiffness of the drive gear, and
- leakage losses
are reduced to an absolute minimum.

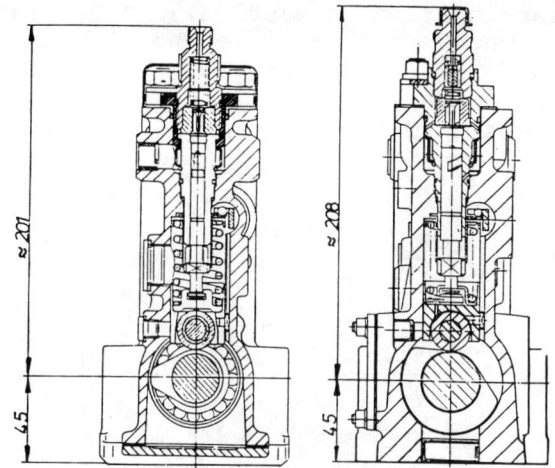

	P1/3000	P 7100
PLUNGER LIFT	10/11 mm	12 mm
BASE CIRCLE	32 mm	34 mm
CAM WIDTH	16 mm	18 mm
PERMISSIBLE PRESSURE PUMP-SIDE FOR Ø 12 mm	750/800 bar	1000 bar

Figure 5 - Comparison between P 1/3000 and P 7100 pumps.

The cross-section of figure 5 and the longitudinal section of figure 6 show the most important differences of the P 7100 in comparison with the P1/3000:
- the increase in the cam lift from 10 mm/11 mm to 12 mm.
- the application of forged flange-type plunger-and-barrel assemblies.
- a camshaft base circle of 34 mm instead of 31 mm.
- a housing which is partly closed in the bottom section, and which is sealed by means of rubber-coated capsules,
- the reinforced intermediate bearing section, in addition to steel bearing shell.
- non-floating bearing principle (cylindrical roller bearings and self-aligning ball bearings).

The basic relationship between the MW and P pumps with regard to design can be seen from the above list.

The FEM method proved here to be extremely useful during optimization of the housing stiffness.

Figure 7 shows the original P 3000 pump housing under load of cylinder 3. The considerable side-area deformation of the housing in the vicinity of the drive can be distinguished at once, showing the need for reinforcement. In addition, the area around the intermediate bearing is particularly at risk.

Considerable improvement of rigidity was achieved by modifications to the housing which involved the change from the version with a bottom cover to one with intermediate webs, and the introduction of a steel intermediate bearing. In addition, a 3 dB(A) lower noise radiation of the pump was measured (at a distance of one-half meter) for this version and proved the effectiveness of the measures taken.

The introduction of the non-floating bearing principle resulted in important improvements in the positional accuracy of the control rack and, therefore, also in the accuracy of the delivered fuel quantity. This bearing version was used from the very beginning for the MW-pump, and features a number of important advantages during assembly. Figure 8 shows the results in the form of a comparison between the two versions of the P-pump. Due to the conversion of

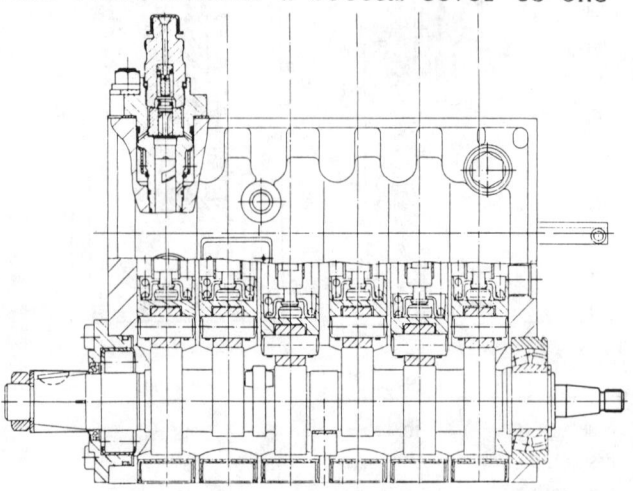

Figure 6 - Longitudinal section of the P 7100 pump.

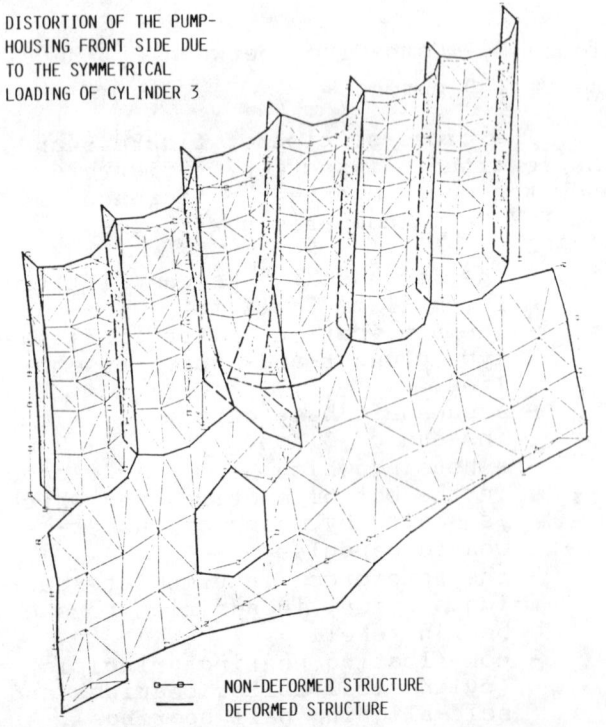

Figure 7 - FEM-analysis of base mounted P 3000.

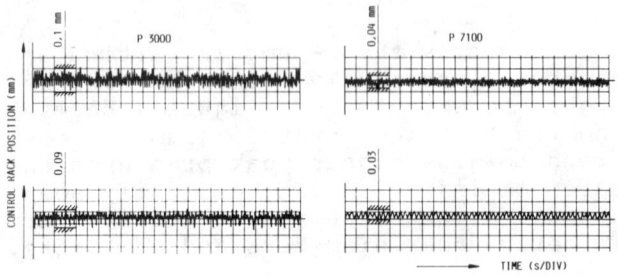

Figure 8 - Stability of control rack travel in relation to the type of camshaft bearing.

the hydraulic peak pressure into an axial force component at the tapered bearing, pulsating camshaft movement is generated during operation and consequently, control-rack dither as well. This has a negative effect on the stability of the deliverd quantity. This phenomena becomes more pronounced with higher temperature because with the tapered bearing design, a temperature-dependent increase in play arises between the aluminim housing and the steel camshaft.

The camshaft oscillation was approximately halved by the introduction of the spherical roller bearing at the governor end. The position of the governor linkage also remains constant independent of the temperature.

By using the camshaft as the inner roller track at the drive end, it was possible to achieve a considerable increase in bearing performance while retaining identical installation dimensions. This has resulted in the P 7100 remaining fully interchangeable with the former P pumps.

CONSTANT-PRESSURE VALVES

The measures dealt with up to now, to generate higher injection pressures by increasing the stroke and the rigidity, as well as the measures implemented to increase stability and durability, do not suffice. Steps must also be taken to ensure the hydraulic stability. This means that the fuel quantity map shows no instabilities and has uniform

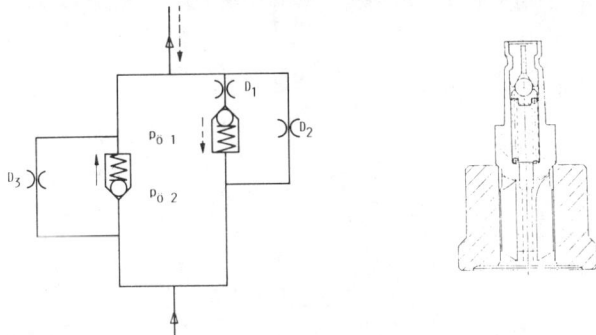

Figure 9 - Constant pressure valve: flow schematic and cross section drawing.

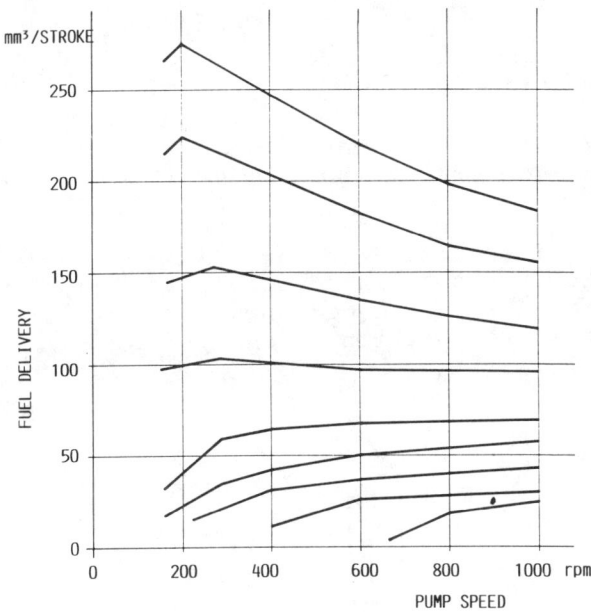

Figure 10 - Hydraulic characteristic of P 7100 with constant pressure valve.

characteristics. In addition, together with the increase in peak pressure at the nozzle, the problems with respect to cavitation and post-injection also increase.

For many years, constant pressure valves have been imperative in the Bosch single cylinder injection pumps for marine diesel engines (pressures over 1000 bar, heavy-oil operation and extremely long service life), and have proved themselves successfully in such applications; it was decided to develop constant-pressure valves for the new increased-power P-pumps.

Figure 9 shows the constructional design of the constant-pressure valve, together with an equivalent schematic circuit.

Figure 10 shows the injected fuel quantity map for a pump with constant-pressure retraction.

Experience has proven that the following advantages can be obtained when using constant-pressure valves:
- stable, uniform injected fuel quantity map.
- freedom from cavitation in the fuel injection lines and at the delivery valve assembly.
- freedom from post-injections.
- increase in the service life of the nozzles as a result of more favorable seat loading.
- positive residual pressure across the total fuel quantity map, particularly in the idle-speed range; this is of considerable advantage for stable governing.
- leak-less nozzles can be employed which operate without return oil, and which are in no danger of jamming.

On the other hand, these advantages are opposed by a number of critical considerations:
- since the retraction process is no longer geometrical, but is now determined by the reflected wave and the opening pressure and throughflow of the constant-pressure valve, the influence exerted by the nozzle and the the fuel injection line on the fuel delivery is larger than is the case with constant-volume valves.
- the injected fuel quantity tolerances can only be completely controlled if the constant-pressure valves are 100 % leak-proof and if the closing behavior of the constant pressure valve is precisely defined. This means that the equivalent orifices D 3 and D 4 in Figure 9 which stand for the untightness of both valves should be practically zero.
- correct design of the throughflow cross-sections and extremely close tolerancing of D1 in Figure 9 is also necessary.
- extremely close definition of the opening pressure tolerance of the constant-pressure valve is necessary. ($p_{ö1}$ in figure 9).

- there exists the tendency for the duration of injection to increase by approximately 0,5 degress camshaft.
- rising fuel characteristic curves in the no-load region have an unfavorable effect on the governor map.
- decreasing full-load curves indicate the necessity for torque control.

The design of this cone/ball valve has proven itself in practice due to its excellent sealing qualities and its closing behavior at the ball seat. As a result of stepless adjustment of the opening pressure which is then followed by final crimping of the seating element, the valve tolerances themselves are also kept low. This design of the constant-pressure valve provides a further possibility for controlling seating and closing behavior due to the additional provision of a small retraction volume at the cone valve, whereby an even more rapid closure of the nozzle needle can be achieved if required.

At the present time, a pilot run of the constant-pressure valves is being subjected to statistic investigations. Large-scale field trials are also being carried out. The use of leakless nozzles provides the engine manufacturer with the chance to economize the injection equipment, as well as being advantageous regarding tolerances during pump-setting.

SPECIAL POSSIBILITIES OF THE PLUNGER HELIX CONFIGURATION

Two important versions of the plunger-and-barrel assembly, for use particularly with emission optimized engines, will be dealt with here briefly in order to demonstrate the versatility of the helix control.

Figure 11 shows the developed view of a plunger for an injection pump incorporating cylinder switch-off at idle for every other cylinder. This is applied in order to reduce HC emissions and fuel consumption. Due to the fact that a divided control rack is not feasible on medium-sized in-line pumps for the purpose of cylinder switch-off, this means that, in the future, increased demand can be expected for the version shown.

A plunger-and-barrel assembly is shown in Figure 12, with upper and lower helixes for the load-dependent retardation of the port closing and, therefore, of the start of injection. This represents a further optimization possibility for adapting to low emission

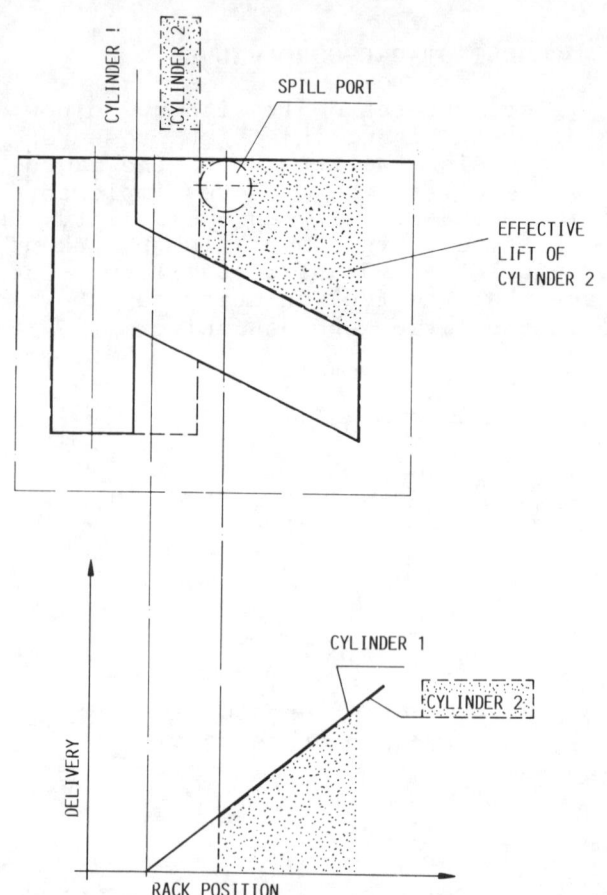

Figure 11 - Schematic overlay of 2 elements with/without cylinder cutoff at idle.

requirements. The three dimensional sketch shows, in addition, a start groove for late beginning of injection at start position of the plunger.

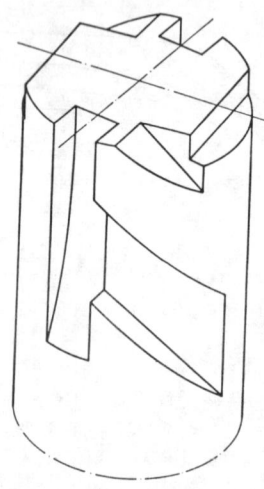

Figure 12 - Plunger layout with double helix for load-dependant timing retard.

NEW SUPPLY-PUMP CONCEPT

A new low-pressure pump has been developed in order to improve the suction height and the delivery quantity. Similar to its predecessors, this is a piston-type pump driven from the injection pump camshaft.

Figure 13 shows a sectional view and the pump schematic. By integrating the suction valve in the working piston, the hydraulic pumping losses of the conventional supply pump were reduced considerably. The delivery stroke at the camshaft was increased at the same time (applies to the MW and P-pumps), with the result that the delivery performance was raised by about 50% to around 220 l/h at 1.5 bar. This means that the temperature distribution in the suction gallery is improved which reduces the cylinder/cylinder deviation.

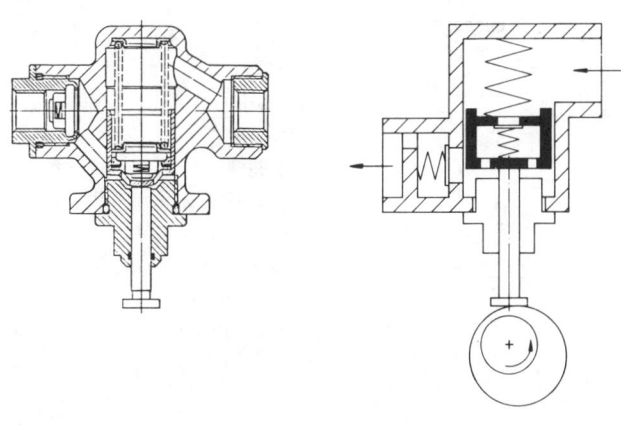

Figure 13 - New feed pump concept: cross section and schematic.

FUTURE OUTLOOK FOR THE INJECTION PUMP

In connection with the state of development currently reached, which permits complete control of pressures between 1000 and 1200 bar, the question arises regarding the limits of the in-line fuel injection pump concept.

The following factors underline the fact that the high power in-line fuel injection pump will continue to be successful as a reliable, versatile, and cost-effective fuel injection system:

- the fuel injection lines present no limits to the system up to pressures of at least 1500 bar, if they are manufactured correctly with regard to material, dimensions and surface finish. This has been proven in pulsation tests.
- the introduction of constant-pressure retraction means that hydraulic stability and power transfer are fully controllable.
- the "active" dead volume of an optimized long-stroke pumpe (i.e., the sum resulting from the dead volumes at both the pump and the nozzle sides), when compared to the metered volume, is similar to that of a unit injector. In addition, it is often the case that pressure can be generated at the nozzle by utilizing the reflection phenomena in the line system.
- a further development step of the long-stroke concept is possible with todays transfer line equipment.

SITUATION REGARDING MECHANICAL (FLYWEIGHT) GOVERNORS:

Notwithstanding the increasing efforts directed towards the introduction of series-production electronic governors, the mechanical governor was not neglected. The very nature of the diesel engine, that is, a highly reliable engine without electrical ignition, means that fully mechanical systems will continue to be needed in the future for a very wide range of applications.

Therefore, the proven RQ, RQV and RSV governors were subjected to an intensive on-going development program in recent years. This served to improve their quality and their functional versatility.

At this point, attention must be drawn to the availability of the following facilities:

- plus/minus torque control concept for the RQ and the RQV governors.
- temperature-dependent starting quantity for all types.
- boost-pressure compensation with precise adjustability.
- key start/shut-off device using add-on electric motor.

ELECTRONIC DIESEL CONTROL (EDC)

With the electronic diesel control (EDC), the mechanical governor, which shifts the control rack depending upon load and rotational speed, is replaced by an electromagnetic actuator with control rack travel feedback. The desired position of the actuator is defined by the microprocessor of the ECU (electronic control unit). The desired position of the rack is calculated by the ECU dependent on the sensor signals.

The current for the actuator is controlled by a closed loop circuit. In contrast, the mechanical governor transforms directly the speed dependent flyweight movement to the guide sleeve and then to the control-rack. The load demand and necessary corrections can be set by the mechanical linkage.

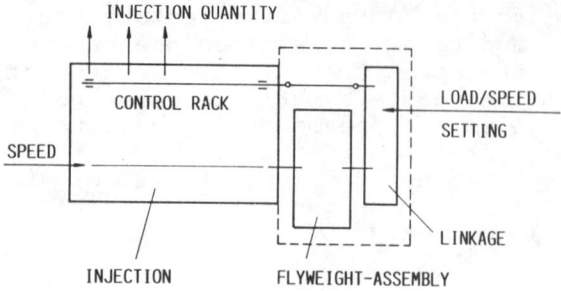

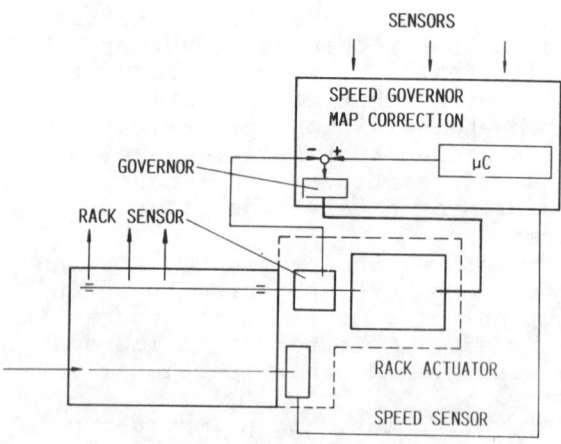

Figure 14 - Schematic comparison of mechanical and electronic governor.

Figure 14 shows the schematic comparison between the mechanical and the electronic governor.

Compared with the direct conversion of flyweight force into control-rack travel as takes place in the mechanical governor, the use of electronic governing at first appears to be the more complicated method. In addition to the most important sensors for rotational speed, control-rack travel and drivers demand, as well as the control unit with position governor and map memory, the electronic governor needs in addition an electromagnetic actuator to shift the control rack.

First of all, what are the advantages of such a system that justify the considerable expenditure that is involved?:

Improved functions and performance	New functions and advantages
Key start/stop	Fuel temp. compensation
Torque curve selection	Air temp. compensation
Driveability characteristics	Cruise control
Zero droop idle control	Vehicle speed limitation
Closed loop timing control	Exhaust temp. limitation
Tamper resistance	EGR control
Smoke control char. (absolute boost pressure)	Surge Damping
	Boost control
Intermediate speed control (droop close to zero)	No mech. linkage from acc. pedal to pump
	Interface capability:
	- Fuel flow output
	- RPM output
	- Diagnostics
	Easy application and change by software
	Pump inventory reduction

Figure 15 shows the block schematic of an electronic diesel control (EDC) with control-rack actuator. The desired-value map for the control-rack travel, and therefore for the injected fuel quantity as a function of rotational speed and drivers demand, is stored in its data memory. A value for the desired control-rack position is calculated in accordance with the signals received from the sensors. The current through the electromagnetic actuator, and therefore the force applied by it, is controlled so that the control-rack assumes a desired position against the force of the return spring. This desired-position signal is generated by the position governor in the ECU which compares the actual control-rack travel signal with the control-rack travel reference signal calculated by the governor. It is also possible to output signals. Dependant of the function, the principle of minimum or maximum value selection is applied. Minimum value selection is

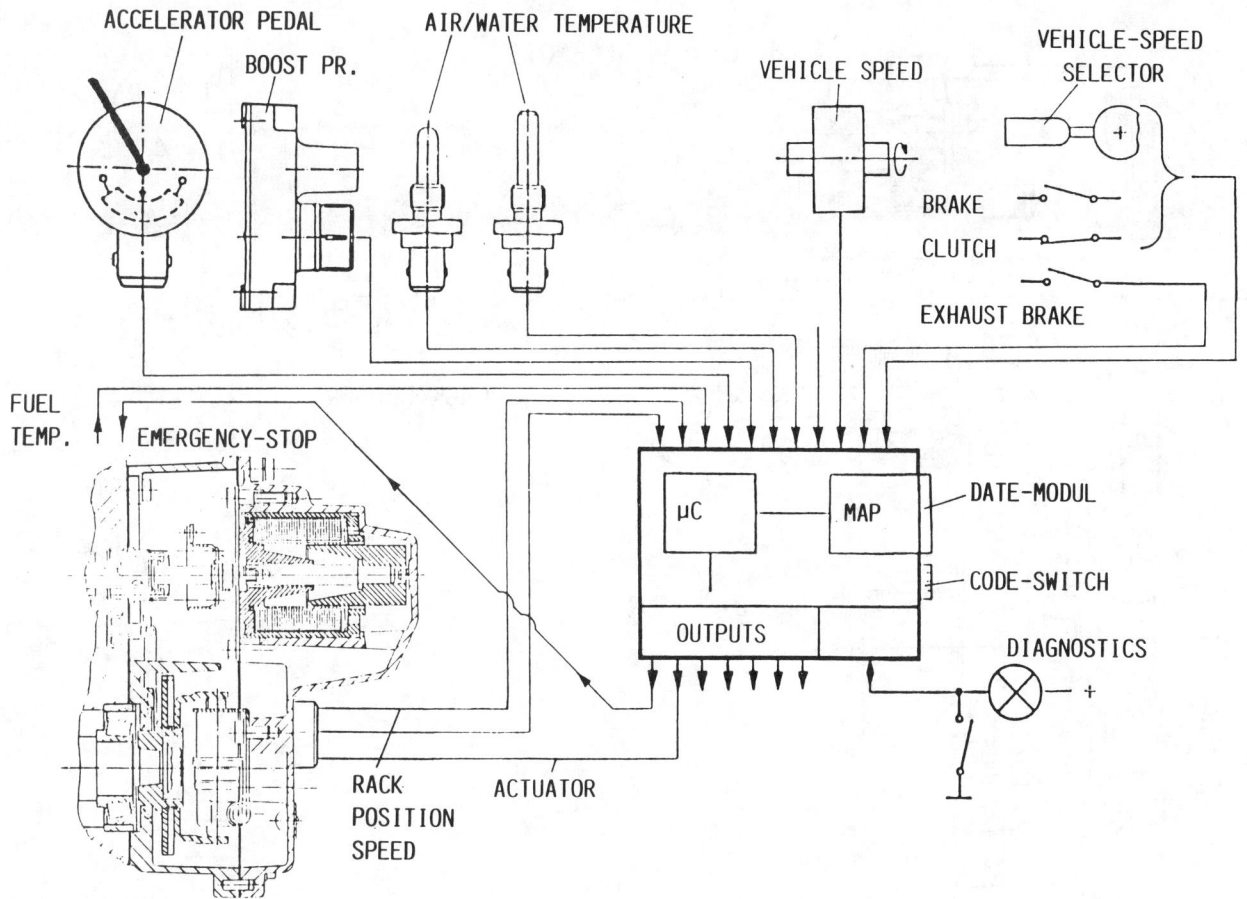

especially for the function of smoke limitation; an example for the maximum value selection is the cruise control.

As microprocessor, a commercial 8-bit module is used. Signal input and output is almost totally hybridized. Electromagnetic interference is largely avoided by use of electronic filters. Special care was excersized to provide high reliability of the ECU and all other components regarding the connectors and the packaging design, due to the harsh environmental conditions of a diesel engine.

Two versions of the rack position actuators on the fuel injection pump were developed:
- a purely solenoid-operated actuator, which can be used with all models of the MW and P-pumps, and
- an electro/hydraulic actuator which was developed specifically in co-operation with the Mercedes Benz Corp., for the high forces which must be applied on injection pumps with more than 8 cylinders, and which are required at extremely cold temperatures.

Figure 15 - EDC: commercial vehicle system.

Within the framework of this report, we now intend to deal in more detail with these two actuators which, in their function as the interface between electronics and mechanics, assume a position of vital importance.

The targets behind the development of these actuators were:
- guaranteed ease of movement even under the most extreme temperature and position conditions, i.e., sufficient power reserve must be available for operation and above all for shut-off at all temperatures.
- simplest mechanical design.
- maximum possible durability at all eletromechanical interfaces, and particularly at the electrical contacts.
- absence of wear.

The aims were set high in order not to endanger the complex new system due to electromechanical problems. The difficulties here lie in the mastery of the comlex boundary conditions which prevail when equipment is mounted on a fuel

23

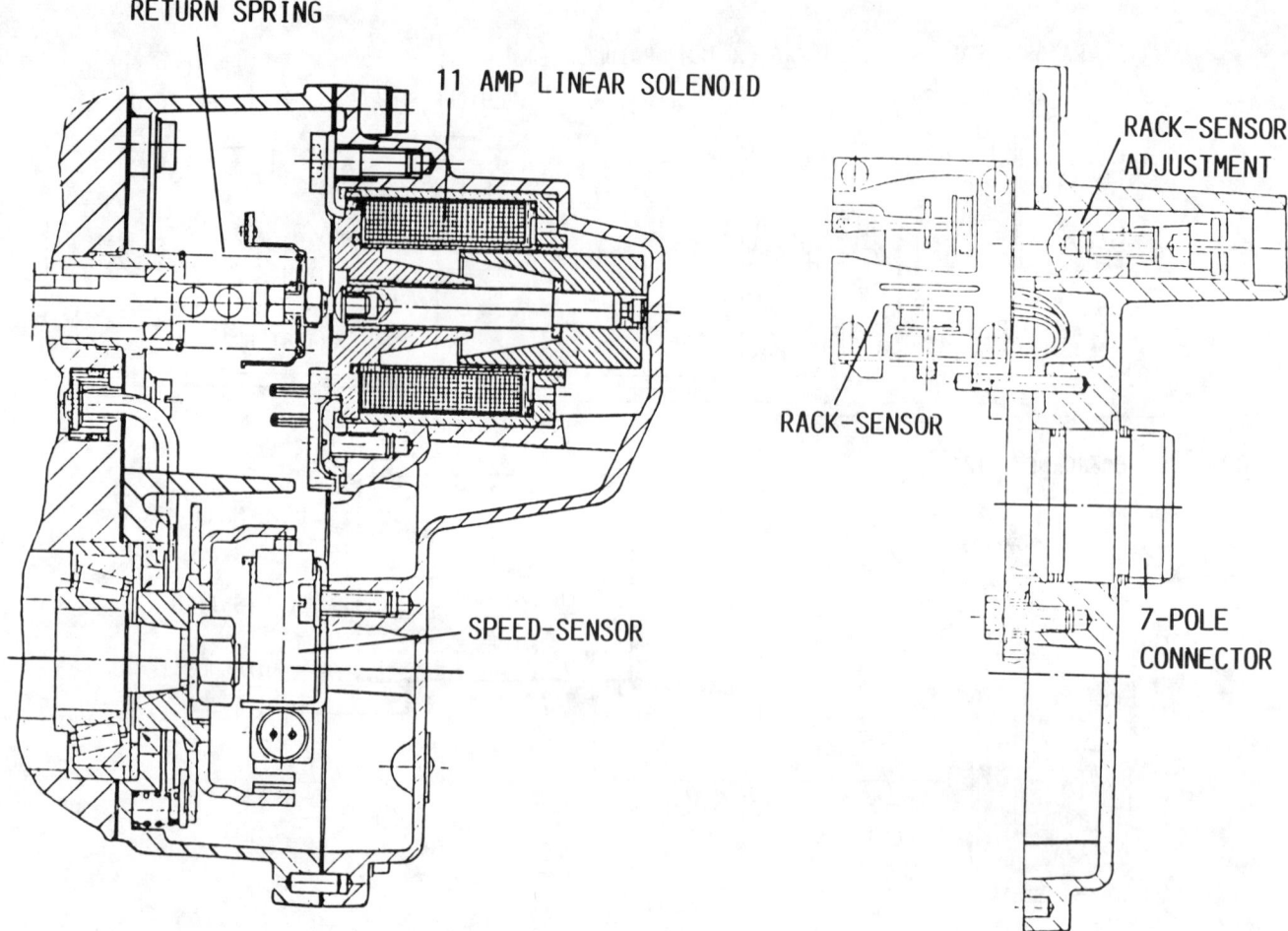

injection pump:
- a service life of more than 10,000 operating hours.
- vibrational acceleration of up to 70 g.
- during actual operation, temperatures between -30°C and +120°C.
- hot soak of up to 150°C.
- difficult environmental conditions due to oil, grease, water, etc.

Figure 16 shows the design of the solenoid actuator intended for the MW and P-pump family.

Here are a number of items regarding the design of the solenoid actuator:
- precision linear solenoid with teflon/bronze bushings fitted coaxially to the control rack.
- return spring acting in the "Stop" direction.
- inductive-type rotational speed sensor and trigger wheel with twice the number of teeth as there are engine cylinders.
- cumulative adjustable, wear-free, eddy-current sensor with induction ring on the control rack, temperature compensated by means of the second coil.

Figure 16 - Electromagnetic actuator for in-line pumps.

- seven-pole, heavy-duty, round-pin plug combined with plastic wire-guide plate, and with encapsulated copper wires.
- vibration-proof wire ends, connected using welding techniques.

In order to ensure good dynamic behavior in both the warm and the cold state, care must be taken that the solenoid is not covered with oil. In this respect, a special version with sealed camshaft and sealed control-rack is available for cases in which the installation position is very unfavorable or when severe specifications must be fulfilled regarding steep terrain performance. Leakage oil is either drained into the engine by means of a special return-oil line, or it is pumped to the camshaft chamber by means of a simple gap-type oil pump, integrated in the governor housing.

24

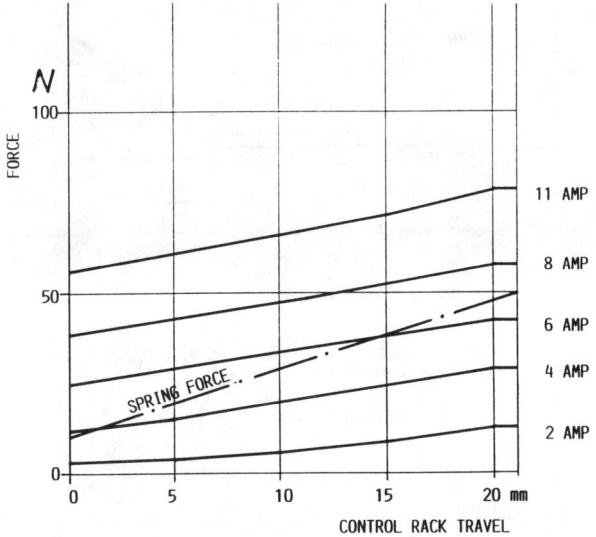

Figure 17 - Force curves of electric actuator.

Figure 17 shows the characteristic curve of travel versus force for the 11 amp solenoid and the return spring. The spring forces realized correspond to those of the mechanical governor.

In Figure 18, the step-response behavior is demonstrated. A mean regulating velocity of 300 mm/s can be realized.

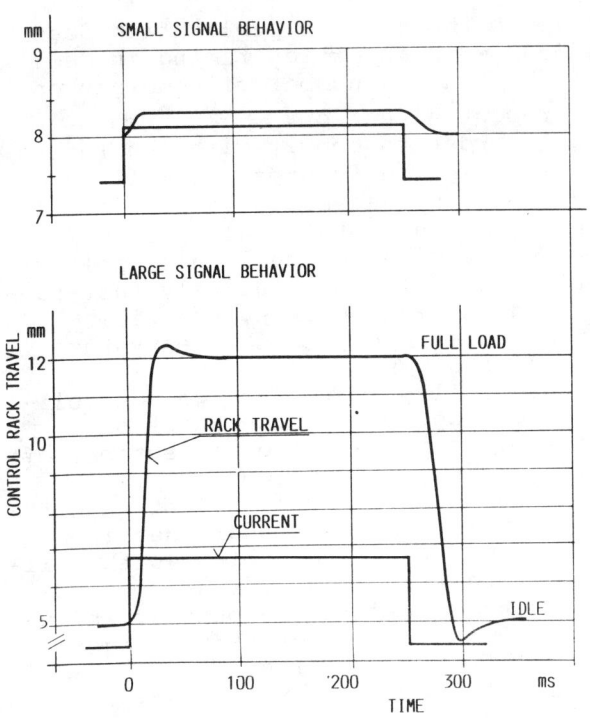

Figure 18 - Step response of electric actuator at 250 pump rpm.

The axial section through the second actuator type, with the electrohydraulic servo function, is shown in Figure 19.

Instead of using a solenoid, here the force against the spring is generated by a hydraulic piston. The magnitude of the servo pressure, and with it the position of the piston, is

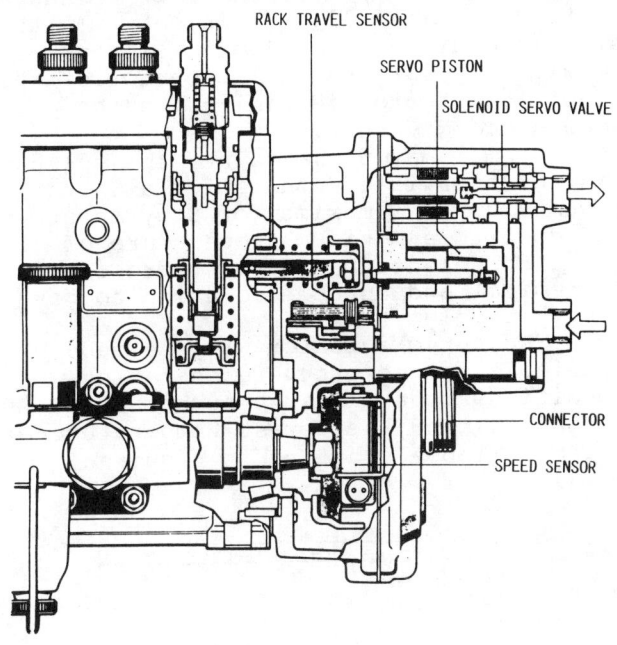

Figure 19 - Cross section of electrohydraulic governor.

Reliable governing is the most important safety requirement when considering a diesel engine. For this reason, ease of movement of the solenoid actuator is a vital factor. Up to the present, about 250 different prototypes were tested during the development phase, and the actuator proved itself under a variety of different test conditions during a total of around 70,000 operating hours without the solenoid jamming. Today, the ease of movement of the solenoid actuator is judged equally as good as that of the mechanical governor. The layout of the solenoid actuator enables, in case of failure of important position governor components, regulated operation without position feedback. By this, a limp-home function is possible. For safety reasons against overspeed, if the microprocessor fails, a redundant overspeed protection will be available. Further more, the position governor is regularly surveyed and, in case of memory integrity loss, the fuel flow into the pump is interrupted by a solenoid valve.

pressure source. This results in the power relationships shown in Figure 20. The forces are more than double those obtained with the solenoid actuator. From Figure 21 it can be seen that a velocity of approx. 100 mm/s is achieved. The slower positioning velocity of hydraulic actuator in the large-signal behavior is caused by delivery limitation of the low-pressure system, but is fully sufficient. In the small-signal behavior, both versions are equal. The large-signal behavior of the electric actuator is similar to the mechanical governor.

From today's viewpoint, the solenoid actuator is adequante for all P-pumps up to a maximum of 8 cylinders and down to a minimum temperature of -25 °C. The solenoid actuator represents an economical and durable solution to governing problem.

The servo actuator will be made available for more severe operating conditions regarding the number of cylinders, and the lowest temperature at which reliable starting is required.

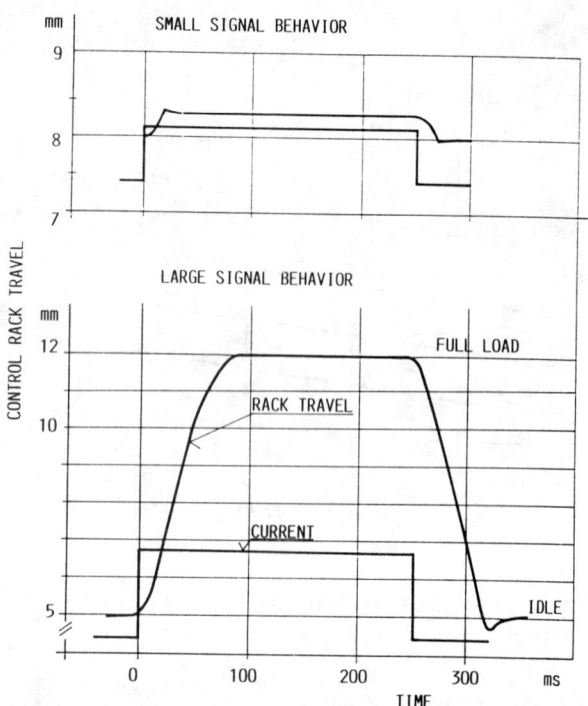

Figure 21 - Step response of electrohydraulic actuator at 250 pump rpm.

START OF INJECTION CONTROL

Even though this publication does not deal with injection timing in detail, here are a number of comments on the present situation.

In order to comply with the future exhaust emission figures, it will more than likely be necessary to have available a load and rotational speed dependent timing device. This applies particularly to the relatively fast-running Medium Duty engines, but will probably also apply to the Heavy Duty class of engines.

Basically, there are two possibilities for controlling the start of delivery, or the start of injection, of an in-line fuel injection pump:
- in the gear train by rotating the primary and secondary sections relative to each other with full torque applied.
- integrated in the pump by influencing the prestroke (plunger lift-to-port closing).

There are a number of possibilities for both groups of solutions. We cannot yet report on the integrated method. One add-on solution would be a pump hydraulic timing device in accordance with Figure 22, the circuit being shown in Figure 23. This timing device was

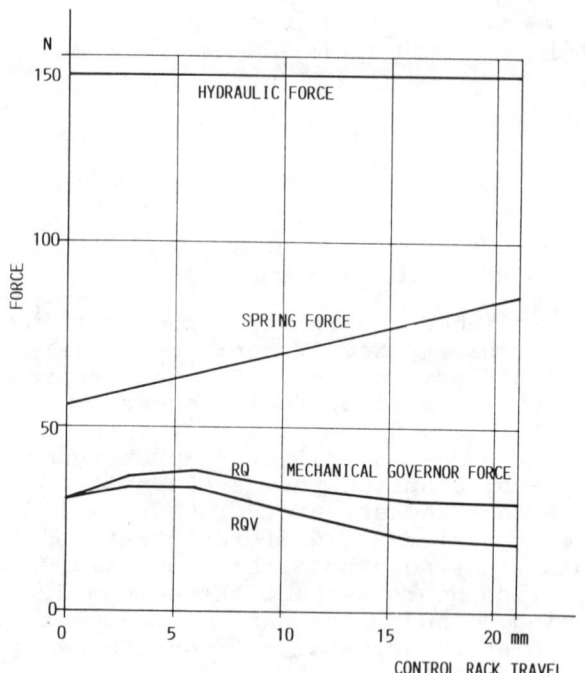

Figure 20 - Force curves of electrohydraulic actuator.

controlled by the solenoid valve mounted above the piston. The remaining components are identical to those of the solenoid type. This applies to the rotational speed sensor, the control rack travel sensor, the contacting principle, and the plug. An electrical fuel supply pump which provides a system pressure of 3 bar is used as the servo developed especially for the MD class of pumps, and has the advantage of fitting well in the in-line pump modular assembly principle. The integrated solutions are only practical when a large proportion of the engines require injection timing advance. It is planned to control the timing-device actuator by means of a special module in the control unit of the electronic diesel control (EDC). Of course, this module can also be a seperate unit by itself.

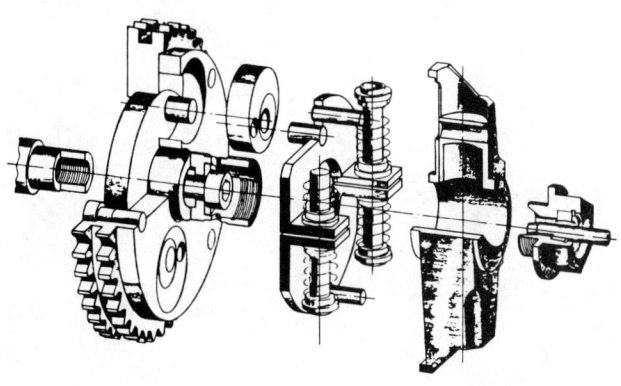

Figure 22 - Cross section of hydraulic timing device.

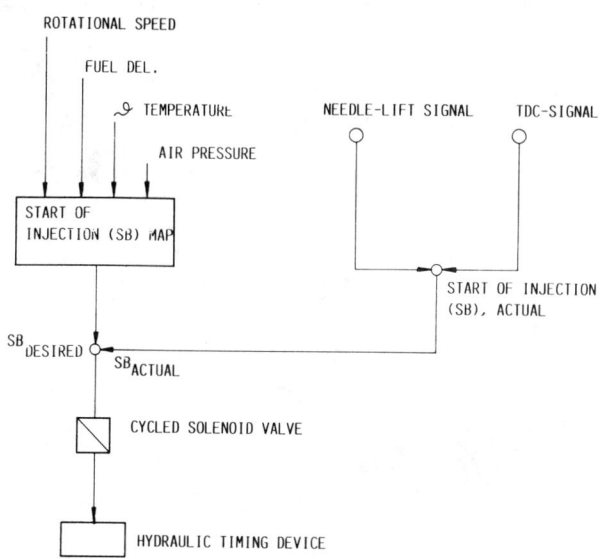

Figure 23 - Schematic of hydraulic timing device.

CONCLUSION

For the stringent and many-sided requirements of the MD and HD class, the further development of the MW/P in-line pumps show themselves to be a flexible solution. Through their performance increase they provide the possibiliy of improvement in engine performance. In spite of worldwide discussions concerning electronic diesel control, the further development of the mechanical governing concept shouldn´t be neglected, because it will continue to be a basic element for many engines and markets.

The introduction of the electronic diesel control (EDC) to the commercial vehicle sector can be expected in the near future. EDC promises considerable refinement of adaptation as well as new functional possibilities. Commercial vehicle applications demand not only functional accuracy but also high degrees of reliability and durability. The target of this paper was to recognize these demands, and the further emphasis was to combine the needs of commercial vehicle diesel engines with the engineering and manufacturing skills which have been successfully applied to electronic control units in the passenger car field.

REFERENCES

1. M. Straubel and R. Schwartz, "The Robert Bosch In-Line Injection Pump (Type "P") for Diesel Engines-Further Development of a Proven Line of Injection Pumps", SAE 780 770.

2. M. Straubel, R. Schwartz, and K. Hummel, "The Robert Bosch In-Line Pump for Diesel Engines, Type MW, Design, Application, and Further Development", SAE 790 901

3. K. Zimmermann, "Elektronischer Regler Fuer Nutzfahrzeug-Dieselmotoren", SAE 845 031.

850172

Development of a Fully Capable Electronic Control System for Diesel Engines

Makoto Shiozaki, Nobuhito Hobo, and Ichiro Akahori
Research & Development Dept.
Nippondenso Co., Ltd.
Kariya City
Aichi-pref. Japan

ABSTRACT

Introduced herein is an electronic control system for controlling a fuel quantity and an injection timing of an in-line fuel injection pump by the use of microcomputers and a servo mechanism for a high precision positioning of a control rack. In the system a new control method of atmospheric pressure compensation is applied. The method maintains engine performance steady under any atmospheric pressure. A digital servo based on the modern control theory is employed as a servo mechanism for positioning the control rack in the fuel quantity control so that capability of the servo system is enhanced.

AS A RESULT OF INCREASINGLY severe requirements in a new model Diesel car, it has become inevitable that certain criteria must be satisfied in relation to problems such as fuel consumption, emissions, drivability, noise and so on. On the other hand, excellent LSI (Large Scale Integrated circuit) or microcomputers which function reliably on a vehicle despite a harsh environment have recently become available owing to significant improvements in electronics technology. An electronic control system for controlling a Diesel engine by means of a microcomputer has already been developed as an outcome of the above two trends (1, 2, 3)*.

In this paper we shall describe an electronically controlled fuel injection system developed in our company and discuss its capability.

---

*Numbers in parentheses designate references at end of paper

ELECTRONIC DIESEL ENGINE CONTROL

The fundamental items proper to an electronically controlled fuel injection system are respectively, fuel injection quantity control and injection timing control, both being important to improvement in fuel consumption, purification of exhaust emissions, enhancement of drivability and attenuation of noise/vibration in the engine.

The characteristics of an electronically controlled Diesel system equipped with a microcomputer are listed below.

EXCELLENT APPLICATION FLEXIBILITY - Any system equipped with a microcomputer is provided with a memory on which various control parameters can be stored. The system can be programmed for the specific numerical values selected by individual engine manufacturers.

EXCELLENT CONTROL ACCURACY AND HIGH RESPONSE - When a microcomputer is used, appropriate response to any variation in the engine environment is rendered possible because of the flexibility in the setting up of the control algorithm, thus enhancing overall control accuracy and response.

REALIZATION OF COMPOUND CONTROL - When a microcomputer is used, one electronic control unit (ECU) can provide the commands necessary to optimize the relationship of the control parameters. The electronically controlled Diesel system described in this paper is able to carry out two kinds of control simultaneously, namely injection quantity control and injection timing control, by means of one ECU. System cost can be reduced by common use of a sensor, and cost/performance of the ECU can be increased by carrying out individual control with one ECU.

OPTIMUM CONTROL - The exhaust emission regulations to which Diesel engines must comply are tending to be more severe each year. Engine manufacturers must satisfy those regulations with minimum fuel consumption compromises. Application of a microcomputer with high operational capability is considered

indispensable in resolving this conflict.
IMPROVEMENT IN SERVICEABILITY - A microcomputer can have an in-built capability to diagnose system troubles by itself. This capability implies the detection of sensors/actuators malfunctions within a system, and speeding up the detection/correction process by appropriate visual or audible signals to the serviceman.
FAIL-SAFE FUNCTION - When a microcomputer is used, it becomes possible - due to the microcomputer's ability to detect sensor trouble - to continue operation of the system without the need to shut the system down. This is done by setting a predetermined value (default value) as the output of the sensor in anticipation of a malfunction.

An electronic control system employing a microcomputer has many advantages, as stated so far, and the essential point is to design a system which makes use of these advantages.

CAPABILITY OF ELECTRONICALLY CONTROLLED FUEL INJECTION SYSTEM

Capability is a parameter which, in addition to other items such as control accuracy and response, should be taken into consideration in constructing an electronically controlled fuel injection system. Capability expresses the strength of a system and signifies stabilized controllability against any variation in ambient conditions.

What kind of system constitution and control algorithm are desirable for enhancement of this capability? As an example, injection timing control following variation in atmospheric pressure, shall be considered.

The optimum injection timing for an engine is determined in an electronically controlled fuel injection system as shown, for example, in Fig. 3-1. The injection timing pattern shown in Fig. 3-1 is considerably more intricate than the pattern of a conventional mechanical timer (Fig. 3-2). The ability to hold an injection timing pattern expressed by a complex curved surface is evidence of one large advantage obtained from electronic control.

How is optimum injection timing for an engine determined? Fig. 3-3 shows the engine operating range divided into sections according to the evaluation factors that are considered most significant. Emissions, fuel consumption and smoke quantity are considered most important in the low engine speed - low/medium load range, medium/high engine speed range and low engine speed - high load range respectively.

Emissions, fuel consumption and smoke characteristics corresponding to variations in injection timing at points in each of the above operating ranges, are displayed in Fig. 3-4. In the low engine speed - low/medium load range, shown by ▢ marks in the figure, injection timing is determined by putting

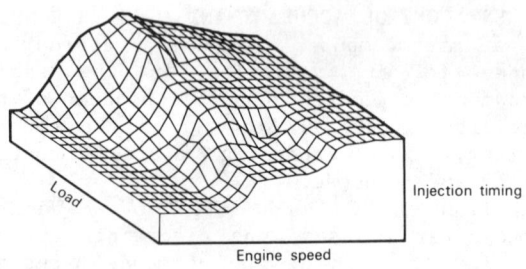

Fig. 3-1 Injection timing pattern of electronic control system

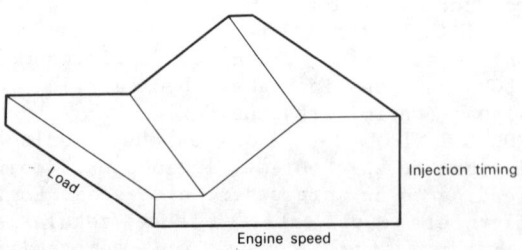

Fig. 3-2 Injection timing pattern of mechanical control system

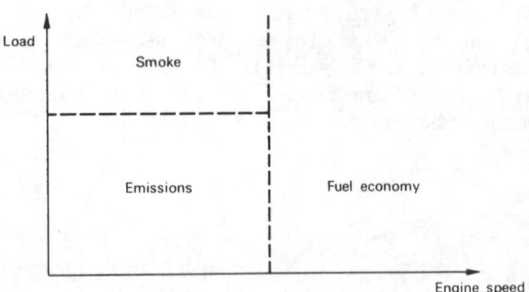

Fig. 3-3 Factors for determining injection timing

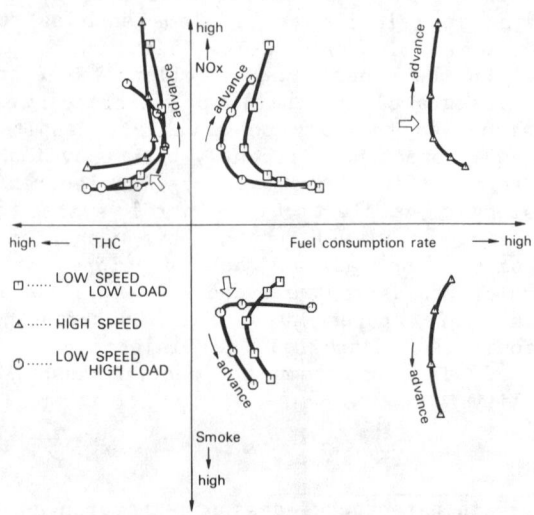

Fig. 3-4 Injection timing vs. emissions, smoke and fuel consumption on various load and speed

emphasis on emissions, so that the optimum injection timing is represented by the point shown by the arrow. In the medium/high engine speed range, shown by △ marks, the optimum injection timing is represented by the point shown by the arrow in the figure, since emphasis is put on fuel consumption. In a similar way, in the low engine speed - high load range shown by the ◑ mark, the optimum injection timing is represented by the point shown by an arrow since emphasis is put on smoke. The optimum injection timing pattern determined from these data is shown in Fig. 3-1.

If an engine with the optimum injection timing pattern determined under standard atmospheric conditions and by the above mentioned procedure, is operated at higher elevations, it will no longer operate with the optimum injection timing, because the optimum injection timing pattern is now different from that under standard atmospheric conditions.

Pressure compensation is necessary so that the engine may be operated at altitude under the same conditions as at sea level. The present compensation process is shown in Fig. 3-5. Compensation consists of making parallel shifts of the injection timing pattern shown in Fig. 3-1 in response to pressure variations. The process of compensation does not cause any change in the form of the injection timing pattern itself. This pressure compensation process is sufficient provided that the optimum injection timing pattern at altitude where the atmospheric pressure is low, is similar to that at sea level. However, the form of the pattern does in fact change. Injection timing patterns at altitude and at sea level are displayed together in Fig. 3-6. The thicker solid lines in the figure represent the injection timing pattern at altitude. As may be observed in Fig. 3-6, the optimum injection timing pattern at altitude is not just a matter of making a parallel shift of the sea level pattern. The process of shifting a pattern in parallel in response to the pressure does not provide adequate capability against atmospheric variations.

The optimum injection timing patterns corresponding to variations in atmospheric pressure, are stored in the ROM (Read Only

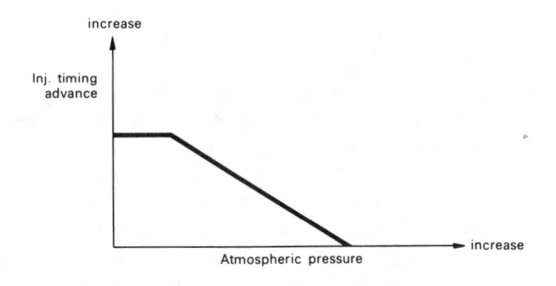

Fig. 3-5 Atmospheric pressure compensation

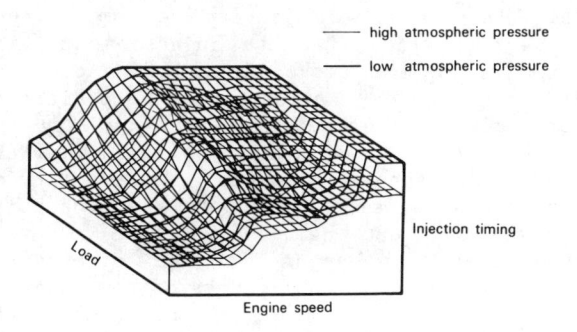

Fig. 3-6 Comparison of injection timing pattern

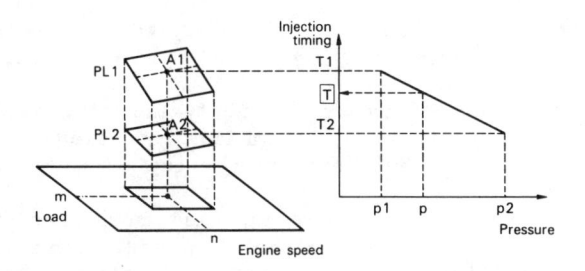

Fig. 3-7 Calculation method for desired injection timing

Memory) of the ECU as a set of numerical data so as to give improved capability. Data groups displaying several injection timing patterns in response to the prevailing atmospheric pressure are stored in the ROM, and the optimum injection timing pattern peculiar to a pressure value sensed by the atmospheric pressure sensor can be obtained by applying linear interpolation to these stored data. This process is shown in Fig. 3-7. Engine speed, load and pressure are n, m and p respectively. Plane PL1 and plane PL2 in the figure represent injection timing patterns under pressure p1 and pressure p2 respectively. Pressure p is more than p1 and less than p2. A1 and A2 represent those points on respective planes where the engine speed is "n", and load is "m". Injection timings T1, T2 specific to planes PL1 and PL2 are calculated by making 3 linear interpolations from the injection timings at 4 points surrounding points A1 and A2. The target injection timing T can be evaluated by applying linear interpolation again to T1 and T2.

The capability of the electronically controlled fuel injection system in response to external changes including atmospheric pressure can be improved by calculating and controlling the injection timing through the above-mentioned process.

Next, injection quantity control following variations of atmospheric pressure will be taken into consideration. Smoke is the target of the evaluation in this case. The "Smoke limited fuel quantity" under standard atmospheric conditions is shown in Fig. 3-8. When the atmospheric pressure drops, the density of air supplied to the engine also drops, thereby resulting in a reduction of air mass. The relative quantity of smoke then increases provided that the injection quantity is similar to the smoke limited fuel quantity under standard atmospheric conditions. Accordingly, it becomes necessary to vary the smoke limited fuel quantity, following the change in atmospheric pressure. The quantity of smoke generated is determined by the proportion of excess air and engine speed, and air flow varies due to changes in the atmospheric pressure. The resultant process of compensation with respect to variations in atmospheric pressure is shown in Fig. 3-9. Coefficient Kp is evaluated from the atmospheric pressure. Compensation with respect to the atmospheric pressure is effected by multiplying the smoke limited fuel quantity Qs under the standard atmospheric conditions shown in Fig. 3-8 by this coefficient.

The capability in correctly matching fuel injection quantity to the ambient conditions can be enhanced by effecting the foregoing compensation.

Although the above-mentioned compensation is to be applied in response to variations in atmospheric pressure, a similar compensation is also applicable in response to variations in intake air temperature.

ELECTRONICALLY CONTROLLED FUEL INJECTION SYSTEM DIAGRAM

The electronically controlled fuel injection system is shown in Fig. 4-1. The electronically controlled fuel injection system comprises various kinds of sensors and switches, an ECU, actuators and indicator. The kinds of sensor used in the electronically controlled fuel injection system are shown in Table 4-1. A starter switch and a diagnosis switch are included.

Table 4-1 Sensors

Sensor	Detection method	Range
Engine speed sensor	Electromagnetic pick-up	40 - 800 Hz
Injection timing sensor	Electromagnetic pick-up	40 - 800 Hz
Control rack position sensor	Variable inductance	12 mm max
Accelerator position sensor	Potentiometer	70° max
Atmospheric pressure sensor	Semiconductor strain gage	66.7 - 266.6 kPa (500 - 2000 mmHg abs)
Air temperature sensor	Thermistor	−20 - 80 °C
Coolant temperature sensor	Thermistor	50 - 120 °C
Oil temperature sensor	Thermistor	−20 - 80 °C

The electronically controlled fuel injection system comprises 2 kinds of actuator. One is a fuel injection quantity actuator and the other is an injection timing actuator.

The governor part of the injection pump, including the fuel injection quantity actuator, is shown in Fig. 4-2. The governor incorporates an injection timing sensor, engine speed detecting gear and a control rack position sensor together with sensor amplifier. The injection quantity actuator consists of a solenoid whose core is directly connected to the control rack.

An injection timing actuator is shown in Fig. 4-3. Major components are a hydraulic piston, a shifter, and an eccentric cam. The hydraulic piston moves in an axial direction due to the effect of hydraulic oil flowing into the actuator. The shifter making contact with the piston moves in a radial direction, following the movement of the piston. The structure is so designed that a phase difference arises between the driving shaft side and the driven shaft side due to the movement of this shifter and the operation of the eccentric cam, thus causing the change of injection timing. The quantity of oil flowing into the actuator is determined from the open/close time of two solenoid valves shown in Fig. 4-1.

Among the components constituting the electronically controlled fuel injection system, the ECU (Electronic Control Unit) plays the most important role. Its appearance and block diagram are shown in Fig. 4-4 and Fig. 4-5 respectively. The ECU comprises 2 CPUs, i.e. a main CPU and a sub CPU, between which data are transmitted via a serial communication interface. In the main CPU, the quantity of fuel supplied to the engine is calculated based upon the signals supplied from various kinds of sensors, and data are transmitted to the sub

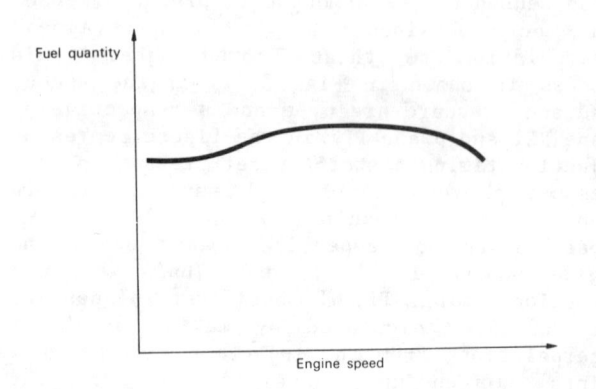

Fig. 3-8 Smoke limited fuel quantity

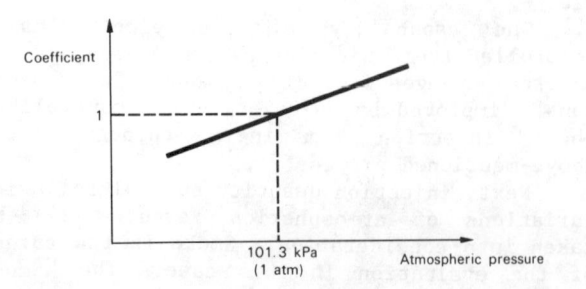

Fig. 3-9 Atmospheric pressure compensation

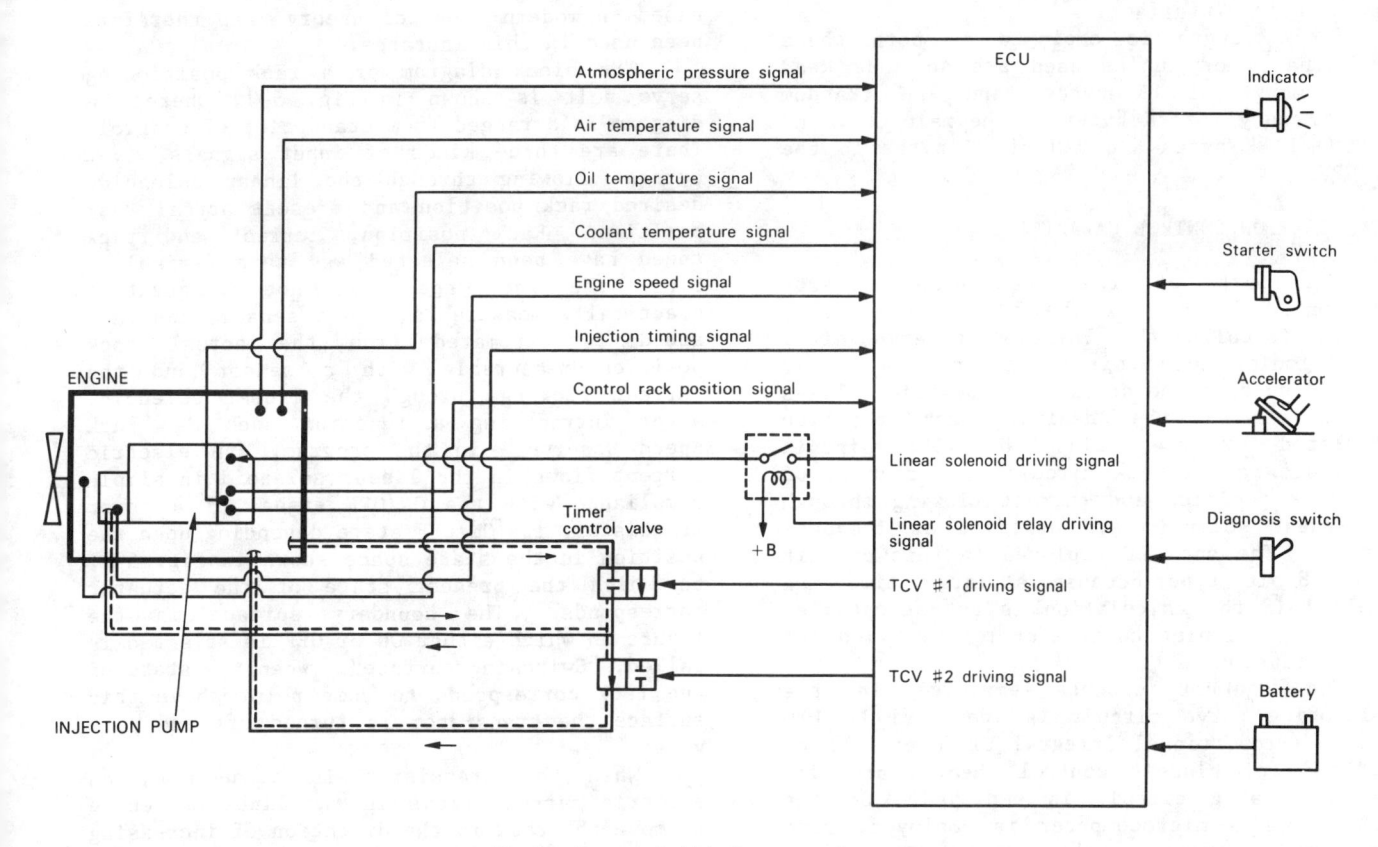

Fig. 4-1 Electronic fuel injection system diagram

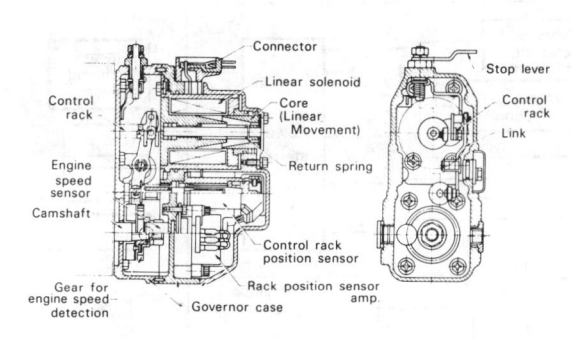

Fig. 4-2 Electronic governor

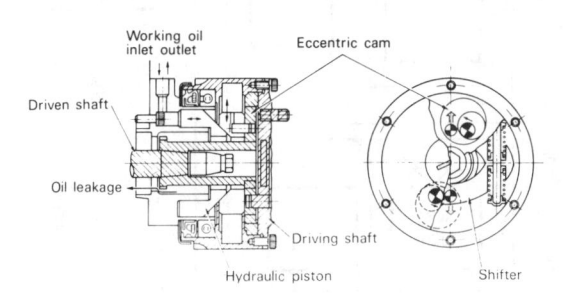

Fig. 4-3 Injection timing actuator

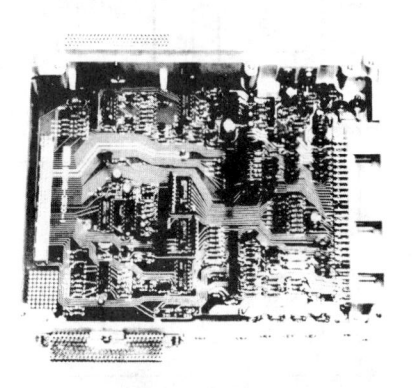

Fig. 4-4 ECU

CPU as the desired rack position. In addition, pulse signals are transmitted from the main CPU to two solenoid valves (TCV; Timing Control Valve) which determine the quantity of hydraulic oil flowing into the injection timing actuator so that the target injection timing obtained by calculation is realized in practice. The sub CPU controls the injection quantity actuator so that the actual rack position coincides with the desired rack position given by the main CPU. Another function of the sub CPU is to control an A/D converter which converts analog signals into digital signals and to transmit the digital signals to the main CPU via serial communication interface.

8-bit length is employed in both these CPUs. The memory units used are an ROM (Read Only Memory) of 8K bytes, and RAM (Random Access Memory) of 2K bytes in the main CPU, and an ROM of 8K bytes and RAM of 2K bytes in the sub CPU.

RACK POSITION CONTROL CAPABILITY

The part of the ECU which effects positioning control of the injection quantity actuator is called the "Positioning Servo Unit". A positioning servo circuit controls a linear solenoid so that the actual rack position always coincides with the desired rack position calculated by the main CPU. The circuit constitutes a feedback circuit, and signals of both rack position and current flowing through the linear solenoid are utilized as feedback signals. The sub CPU employed in this circuit is an 8-bit type because a large time lag involved in the calculation is not acceptable, and the positioning control must be carried out in real time.

The control process employed in the positioning servo circuit is identical to PID control (Proportional Integral Differential) on the basis of classic control theory, provided that an analog circuit is employed. On the contrary if a microcomputer is employed, some new control process other than PID control becomes available. A new control process based on modern control theory has therefore been used in this instance.

The block diagram of a rack positioning servo unit is shown in Fig. 5-1, where the diagram is arranged from standpoint of control. There are three kinds of input signals, i.e. current flowing through the linear solenoid, desired rack position and present actual rack position. Rack position, current and rack speed have been selected as state variables. Since the rack speed is not a quantity practically measurable with a sensor, the rack speed is estimated from the actual rack position measurable with a sensor and the current flowing through the linear solenoid, after introducing a function such as "Rack Speed Observer" in the program. An electric current flows in the linear solenoid in simple compliance with the ON/OFF signal of a power transistor, its ON/OFF state depending upon the position in the state space shown in Fig. 5-1, to which the present state of the actuator corresponds. The boundary surface in the figure on which either ON or OFF is selected is called, "Switching surface". When the state of the rack corresponds to some point above this surface the transistor is turned off and vice versa.

When the transistor is turned on, an electric current flows in the linear solenoid to move the rack in the direction of increasing fuel quantity, while the state of the rack shifts within the state space in Fig. 5-1.

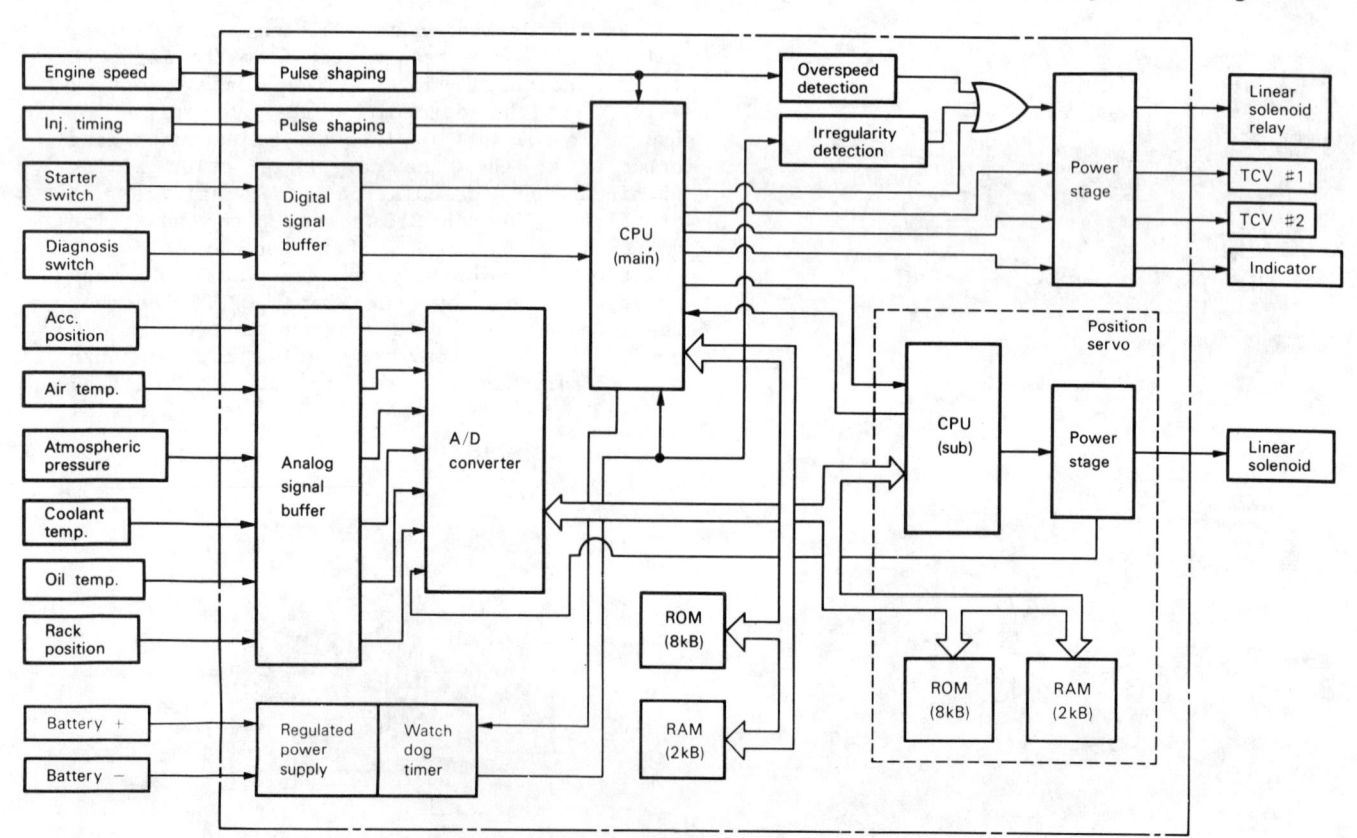

Fig. 4-5 ECU block diagram

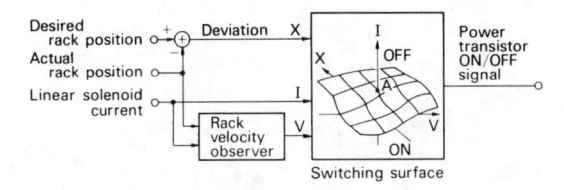

Fig. 5-1 Rack position control

When the state of rack moves into the space on the other side of the switching surface, the transistor is turned off, changing the state of the rack again. Repetition of this ON/OFF action of the transistor forces the state of the rack to move toward point A on the switching surface. Since point A is a point where the deviation of the actual rack position from the desired rack position is zero, bringing the state of the rack to point A is equivalent to controlling the positioning of the rack. The rack is quickly positioned by this process. It should be noted that configuration of the switching surface is so determined as to minimize the settling time (4).

The capability of the positioning servo can be improved by making the rack positioning servo a digital servo circuit in which a microcomputer is employed. Now suppose that the ambient temperature drops, the viscosity of oil in the pump increases, thereby increasing the sliding resistance of the rack. In such a case, the gain on the servo circuit side must be raised in order to secure a high grade overall response, since the response of the actuator part is delayed. When the temperature rises, on the other hand, the gain on the circuit side must be lowered because the response of the actuator part is faster on account of decreased viscosity of the oil. The microcomputer can satisfy the above mentioned requirements very easily. As shown in Fig. 5-2, switching surface data for each temperature level are stored in the ROM, so that the wave form of the response of the rack becomes identical to the wave form at normal temperature. The optimum switching surface is generated from the switching surface data stored in the ROM, based upon the oil temperature obtained by the oil temperature sensor. The same wave form can be obtained, even if the ambient temperature changes, by controlling rack position by means of this switching surface. The capability of the rack positioning system against temperature variation has therefore been improved. Step response wave forms for respective temperatures are shown in Fig. 5-3. As may be seen from this figure, the step wave form hardly changes even if the temperature changes from 20°C up to 80°C, thus ensuring high capability of the rack positioning servo.

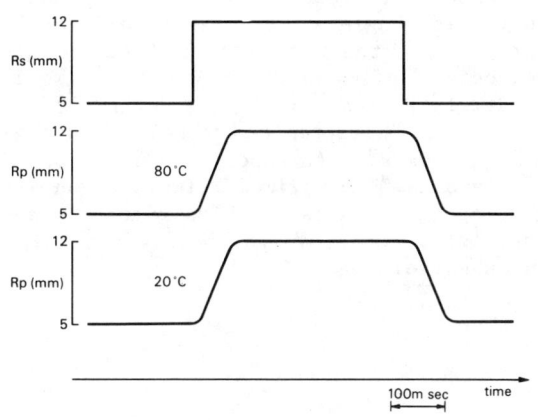

Fig. 5-3 Rack position response

CONCLUSIONS

An electronically controlled fuel injection system employing a microcomputer has been developed for in-line fuel injection pumps. The control functions are related to both fuel injection quantity and timing.

A system more capable of responding to atmospheric variations than the conventional system could be achieved by preparing several optimum injection timing maps with respect to atmospheric pressure. This system is rendered capable also as regards injection quantity control, by making compensation with respect to atmospheric pressure. This concept can be further applied to other types of external effects (temperature).

The positioning servo for the linear solenoid which acts as a fuel injection quantity actuator has been achieved in the form of a digital servo including a microcomputer. A step response of the rack characterized by quick response and small overshoot can be achieved through application of this digital servo. Variations in response wave form caused by variation of oil temperature can also be closely controlled, and thus a positioning servo with high capability against any oil temperature change can be practically realized.

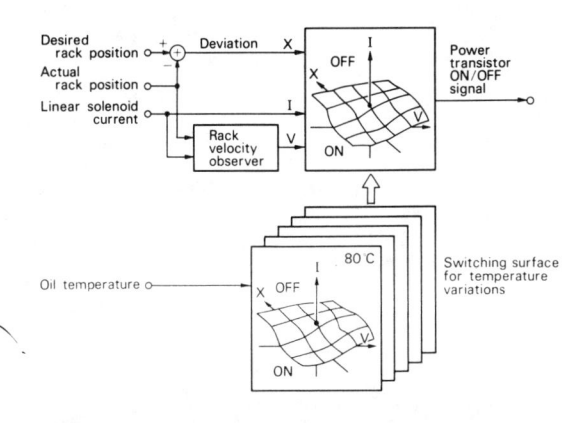

Fig. 5-2 Improvement of capability in rack position control

ACKNOWLEDGEMENT

The authors wish to express their appreciation to Mr. Toshihiko Igashira, Nippon Soken Inc. and to Mr. Hideya Fujisawa, Control Development Dept. of NIPPONDENSO CO., LTD. for their cooperation in this development.

REFERENCES

(1) K. Kageyama, K. Okamoto, "Current Status and Trend of Actuators and Sensors for Electronic Control of Diesel Engines", Journal of the Society of Automotive Engineers of Japan, Vol.38, No.2, pp172-180, February 1984, (in Japanese).
(2) H. Nakao, S. Yamaguchi, "Electronically Controlled Engines for Passenger Cars", Journal of the Society of Automotive Engineers of Japan, Vol.38, No.2, pp195-201, February 1984, (in Japanese).
(3) H. Eisele, "Electronic Control of Diesel Passenger Cars", SAE Paper No.820449.
(4) T. Kitamori, "The Maximum Principle and its Calculating Methods", Instrument and Control Engineers, Vol.8, No.7, July 1969, (in Japanese).

850173

Caterpillar 3406 PEEC (Programmable Electronic Engine Control)

Michael E. Moncelle and G. Clark Fortune
Caterpillar Tractor Co.

Abstract

Electronic engine controls are being developed for heavy duty diesel truck engines to improve performance and fuel consumption at the current and future gaseous emission requirements. Caterpillar is developing a programmable electronic engine control (PEEC) for the 3406 Truck Engine. The in-line 6 cylinder engine currently covers the power range from 255 to 400 hp (190 to 298 kW). The PEEC system has been designed specifically to work with the new fuel system recently introduced on the 3406B Engine (1)[*]. This paper discusses the design, functions and performance aspects of the 3406 PEEC Engine.

[*]Numbers in parentheses designate references at end of paper.

Objectives

In early 1978 it was determined that a new fuel system with higher pressure and faster injection rates, electronic engine controls, and other engine developments were required to improve fuel consumption and performance of the 3406 Engine while meeting future gaseous and noise emissions regulations.

The most common method of reducing diesel engine NO_x emissions is to retard injection timing, figure 1. Fuel consumption and particulates increase as NO_x emission decrease. Engine response or driveability also suffers as emission output levels are reduced. Therefore the following engine performance development goals were established for the 3406 PEEC Engine.

1. Meet all future gaseous and noise emission requirements.
2. Minimize the fuel consumption penalties resulting from stringent emission requirements.
3. Improve engine response for better driveability.

From a user's viewpoint electronics provide improved profitability by offering reduced operating costs, better equipment control, and by providing information for decision making not previously available to the driver and/or management. The customer acceptance goals for the 3406 PEEC system are:

4. Provide a good return on investment to the end user.
5. Reduce operating costs.
6. Provide added driver and management features.
7. Provide reliability equal to or better than today's mechanically controlled engines.

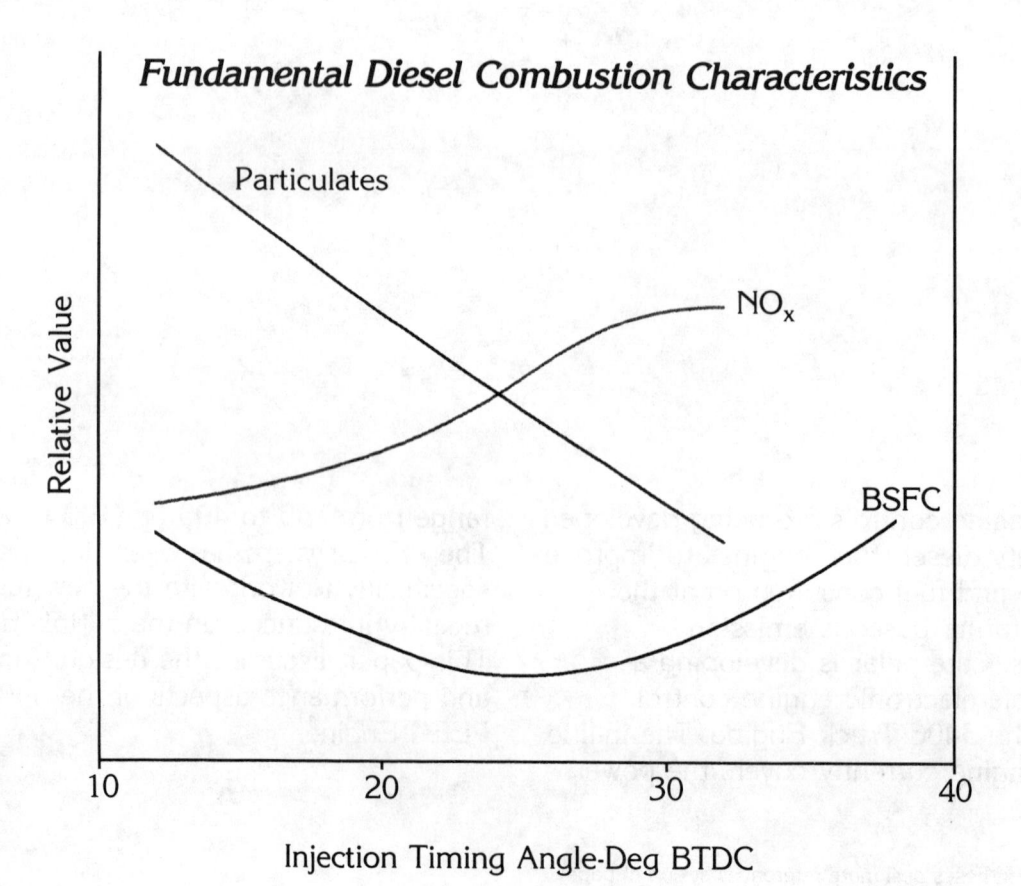

Figure 1

Control Functions

There are two basic types of electronic systems applicable to on-highway trucks: control systems and monitoring systems. Control systems perform an active role in the operation of the vehicle by regulating various processes while monitoring systems evaluate the processes for display to the operator or to be recorded for future analysis.

Caterpillar's approach to electronics for the 3406 has been to develop an engine control that meets the performance development goals and satisfies the customers demand for value. Electronic engine controls have been developed first because of the greatest universal payback to the customer. Additional monitoring features were incorporated in the PEEC system design when minimal cost and reliability impact could be achieved. The following functions will be performed by the 3406 PEEC system. The functions have been divided into the two major areas of engine and vehicle related functions.

Engine Functions

1. *Electronic governing*
   A full speed range electronic governor is used. The governor functions like the Caterpillar mechanical governor in the mid operating range but includes the special features of isochronous low idle and the elimination of governor overrun.
2. *Fuel air ratio control*
   The PEEC system has full authority over engine fuel delivery. The mechanical fuel air ratio control is eliminated. Smoke limiting and engine acceleration rates for noise reduction are more accurately controlled electronically.
3. *Torque rise shaping*
   The engine torque rise can be tailored to limit peak torque or match the engine's performance to the transmission for good driveability. The engine output can be programmed to provide constant horsepower over a large engine speed range.
4. *Altitude and power compensation*
   The PEEC system adjusts engine power and torque rise to compensate for low inlet air pressure resulting from running the engine at high altitudes or with plugged air cleaners.
5. *Injection timing control*
   Injection timing is varied as a function of engine operating conditions to optimize engine performance for emission, noise, fuel consumption and driveability.
6. *System diagnostics*
   The PEEC system performs its own self-health tests to insure that all the PEEC system components are functioning properly.

Vehicle Functions

7. *Cruise control*
   Vehicle speed control is performed by the PEEC system. One of two types of cruise control strategies selected: conventional isochronous vehicle speed control or economy cruise mode control.
8. *Road speed limiting*
   Road speed limiting allows implementation of a gear-fast-run-slow truck specifications to further improve fuel consumption while limiting top vehicle speed.
9. *PTO governor*
   The cruise control functions as a PTO (power take off) governor when the vehicle is stationary. The isochronous engine speed control is even satisfactory for PTO driven auxiliary electric generators.

Control Functions

10. *Data link*
An SAE/ATA compatible data link is provided for communicating engine information to other vehicle electronic systems and interfacing with Caterpillar service tools. The data link can be used to interface with fleet management systems such as trip recorders or electronic dashboards.

System Description

The 3406 PEEC system was designed for the new fuel system developed for the 3406B Engine which delivers fuel at injection pressures of 103 MPa (15,000 psi) while utilizing Caterpillar's inward opening non-leakoff fuel nozzles. The PEEC engine, figure 2, is very similar to the 3406B Engine.

3406 PEEC Engine

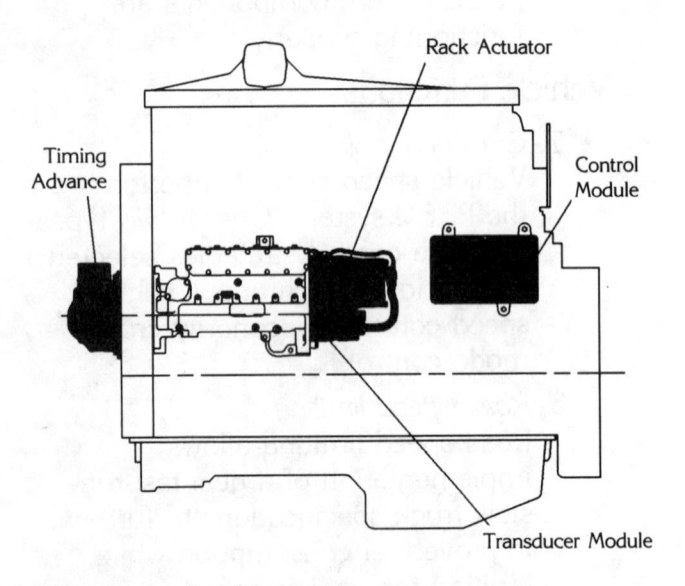

Figure 2

The mechanical speed sensitive timing advance unit has been replaced with an electronically actuated timing advance unit. An electronic rack actuator package which includes a shutdown solenoid replaces the mechanical dashpot governor. A pressure transducer module is mounted on the fuel pump below the rack actuator.

An engine wiring harness connects the timing advance unit, rack actuator and transducer module to the engine mounted, fuel cooled control module. The control module is isolation mounted to the engine block to minimize vibration problems. Fuel from the transfer pump is routed through the control module, then on to the secondary fuel filter prior to entering the fuel pump gallery.

Two connectors on the control module interface the PEEC engine to the truck wiring harness. One connector is for the system data link and the second connector provides power to the control module and signals from the cab and chassis mounted sensors.

A cab mounted throttle position sensor is used to eliminate the troublesome mechanical linkage between the engine and the operator's foot. The cruise control switches (on/off, set/resume, clutch, and brake) are supplied by the truck manufacturer. A cab warning light is provided to indicate a PEEC system malfunction. The cab warning light can be used with a simple service tool to give a flashing coded display of the fault mode, figure 3.

PEEC Component Locations

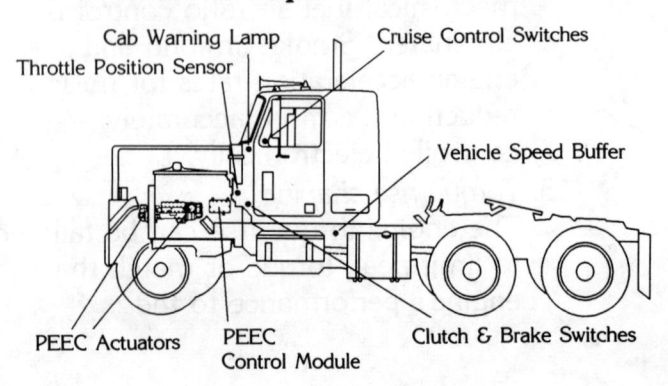

Figure 3

System Description

A Caterpillar supplied vehicle speed buffer is used to send the magnetic pickup generated signal to the PEEC control module and to the truck speedometer.

Timing Advance Unit

The PEEC timing advance unit replaces the mechanical speed sensitive variable timing drive. The engine performance objectives required changing the hydraulic/spring servo mechanism to a double acting hydraulic servo. The double acting hydraulic servo provides the increased frequency response characteristics needed for the PEEC timing control strategies, figures 4 and 5.

The helix angle on the fuel pump camshaft has been increased from 10 to 15 degrees. The straight spline on the drive carrier was replaced with a helical spline to increase the total timing advance range from 9 to 25 crankshaft degrees.

The spring and flyweights are replaced with an electronically actuated brushless torque motor (BTM). The BTM is a special designed rotary proportional solenoid. A linear potentiometer is used for accurate feedback control of the timing advance. Buffer circuitry integrated into the sensor protects the signal from harness faults, electromagnetic interference, inadvertent misconnection, and parasitic harness losses. The timing advance can be adjusted on a running engine to calibrate out mechanical static timing tolerances.

The BTM is mounted toward the inside of the engine and the timing advance feedback sensor is located on top of the actuator housing for easy servicing. The feedback sensor and connectors are protected from mechanical abuse by an access cover.

PEEC Timing Advance Unit

Figure 4

PEEC Timing Advance Actuator and Feedback Sensor

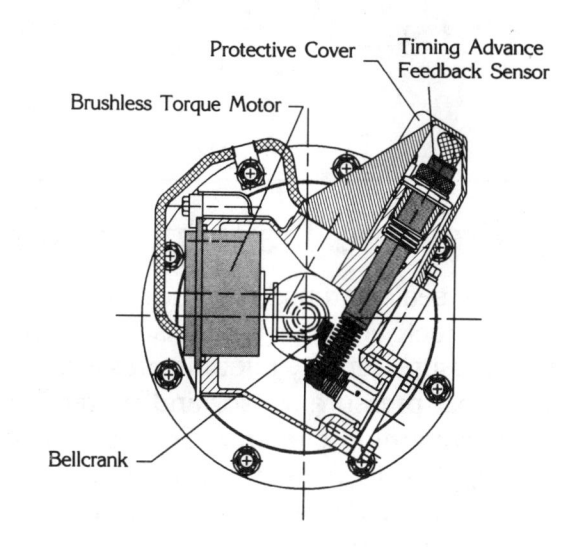

Figure 5

System Description

Rack Actuator

The mechanical flyweight governor package is replaced with an electronic actuator and transducer module. A double acting hydraulic servo provides the mechanical muscle for moving the fuel rack. The BTM actuator for the rack is interchangeable with the BTM used for the timing advance. The BTM is spring biased towards fuel shutoff, figure 6.

A linear position sensor similar to the timing advance feedback sensor is used for closed loop control of rack position. The rack position sensor is magnetically coupled to the rack.

A separate energized-to-run engine shutoff solenoid is controlled by PEEC. The shutoff solenoid provides redundant shut down capability if the BTM is unable to move the rack to the shutoff fuel position. A manually operated mechanical shutoff and shutoff solenoid override is provided, figure 7.

The PEEC engine speed sensor is located inside the rack actuator housing to deter tampering. A powered magnetic pickup is used to detect engine cranking speeds down to 25 rpm. The engine speed sensor is triggered by radial slots on the fuel pump camshaft retainer.

PEEC Rack Servo

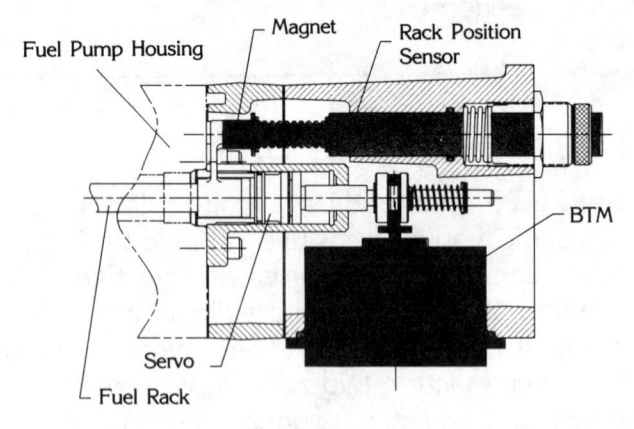

Figure 6

PEEC Rack Actuator and Transducer Module

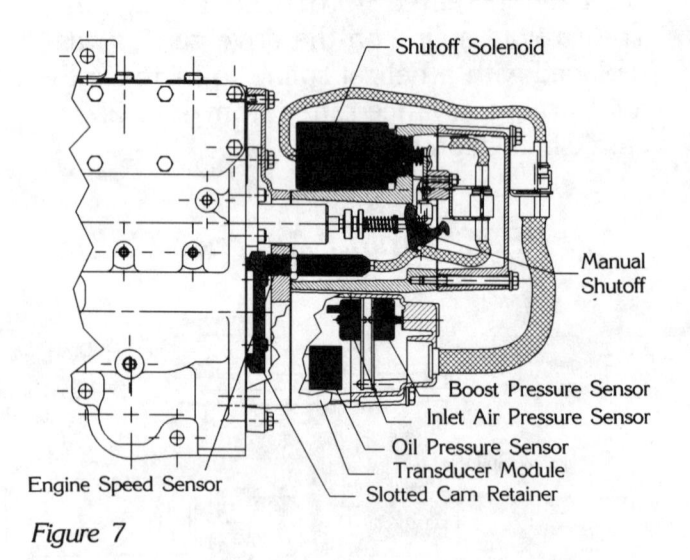

Figure 7

Pressure Transducer Module

The sealed transducer module mounted below the rack actuator contains an engine oil pressure sensor, inlet air pressure sensor, a boost pressure sensor and protective signal conditioning circuitry. Engine oil pressure is supplied to the rack servo and oil pressure sensor by oil passages in the rack actuator center housing. The ceramic capacitive oil pressure transducer is used to limit engine speed and power output if low oil pressure occurs. Boost pressure and air pressure upstream of the turbocharger are routed to the transducer module. The boost and atmospheric pressure sensors use silicon strain gage technology. The rack position and engine speed signals are passed through the transducer module to minimize external connections.

Control Module

The control module consists of two major components; the main control and a ratings personality module. The ratings personality module contains all the engine performance and certification information along with the control software. The main control module contains an electrically erasable programmable read only memory (EEPROM) to store customer specified parameters, figure 8.

The ratings personality module contains the timing, fuel air ratio and rated rack control maps for a particular ratings group that utilizes common engine components. Any engine rating from a number of available ratings in that group can be selected through the data link by using a Caterpillar service tool. The non-volatile memory internal to the control module is used to identify the selected rating. The non-volatile memory contains a personality module identification code to deter unauthorized tampering or switching of rating personality modules. The EEPROM is also used to store other pertinent manufacturing information and customer specified parameters such as the road speed limit, cruise control limit, drive shaft to vehicle speed calibration, owner specified identification and owner entry passwords.

Particular attention has been paid to the heat transfer from the major power handling devices to the cooling medium. Fuel cooling was chosen because of its low nominal temperature and its guaranteed flow during engine operation. The fuel cooling plate for the control module is a one piece die cast aluminum housing. Manifolds on top and bottom of the control module route fuel from the transfer pump through the cooling plate. The package has demonstrated good thermal efficiency under the worst case conditions of power dissipation, fuel temperature, ambient temperature and minimum fuel flow, figure 9.

PEEC Memory Configuration

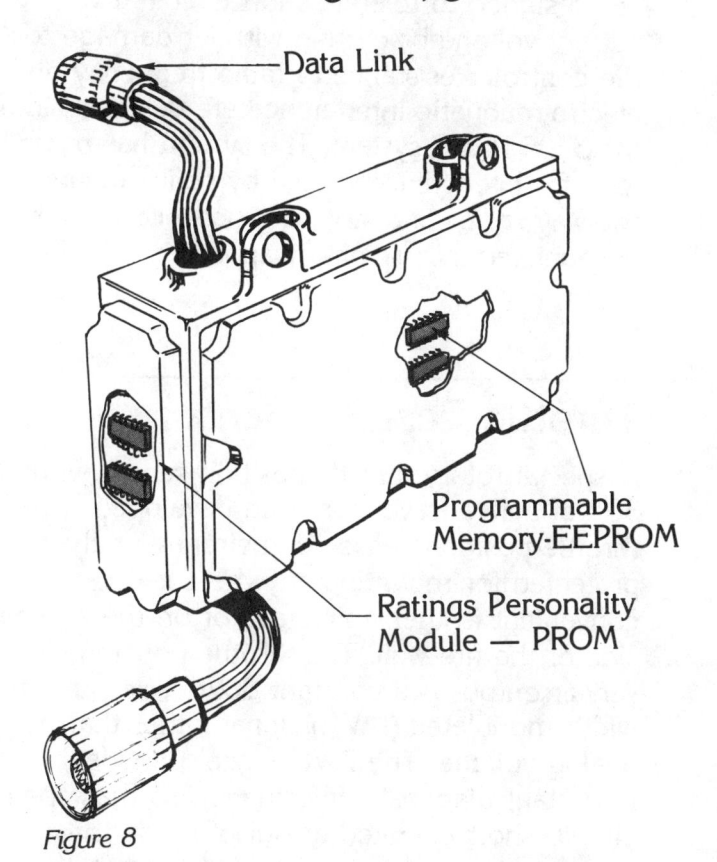

- Data Link
- Programmable Memory-EEPROM
- Ratings Personality Module — PROM

Figure 8

3406 PEEC Control Module

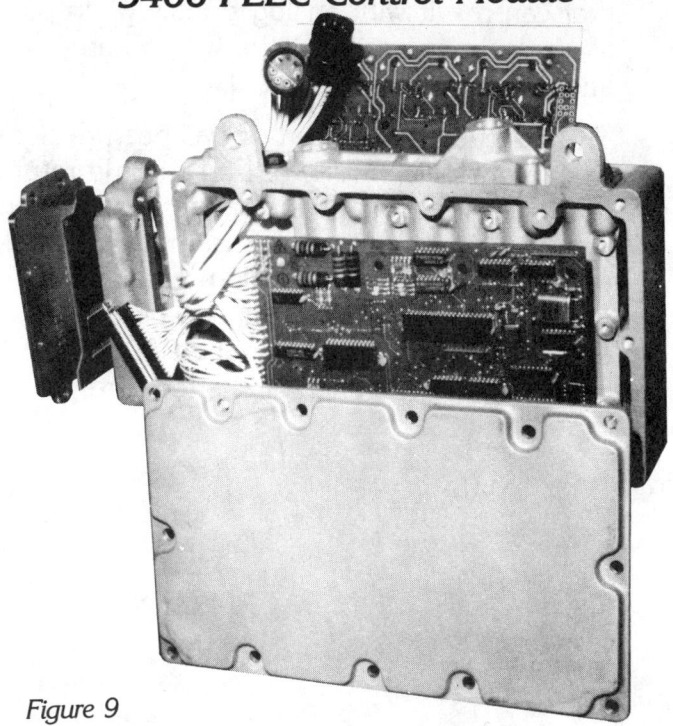

Figure 9

System Description

All inputs and outputs to the control module are designed to tolerate short circuits to battery voltage or ground without damage to the control. Resistance to radio frequency and electro-magnetic interference are designed into the 3406 PEEC system. The system has passed tests for interference caused by police radars, two-way radios and switching noise commonly encountered in on-highway applications.

The control module power supply provides electrical power to all engine mounted sensors and actuators as well as the internal logic. Reverse polarity protection and resistance to vehicle power system voltage transients has been designed into the control. Voltage transients due to alternator load dump and solenoid current decay are more severe in truck applications than in the automotive industry due to the higher currents and larger components used in truck systems.

Throttle Position Sensor

A special rotary throttle position sensor with 30° of active travel has been developed. The throttle position sensor is environmentally protected for mounting anywhere it is convenient in the vehicle cab or on the engine side of the fire wall. The throttle position sensor output is a constant duty cycle pulse width modulated (PWM) signal rather than an analog voltage. The PWM signal provides important diagnostic information so that open circuit, short circuited or out of range throttle position signals can be detected accurately. The PWM signal overcomes the serious errors that can result with analog signals when pin-to-pin leakage occurs in the wiring harness and/or connectors, figure 10.

The throttle position sensor has been designed to comply with FMVSS124 for throttle return under environmental temperature extremes and with part of the throttle linkage missing. Two return springs for the throttle position sensor are located behind the mounting disc. The engine returns to low idle if the PWM signal is invalid due to a broken or shorted wire.

Throttle Position Sensor

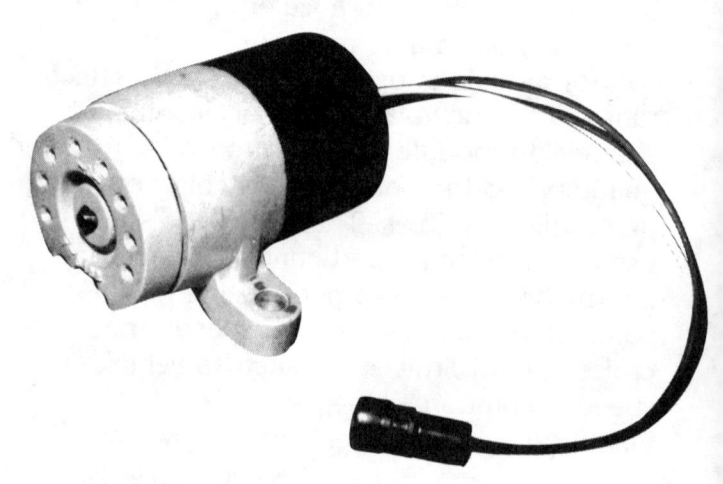

Figure 10

Vehicle Speed Buffer

A buffer circuit is used to amplify and wave shape the output of the magnetic vehicle speed sensor. The vehicle speed buffer prevents overloading of the magnetic pickup when multiple devices need to measure vehicle speed. The conditioned signal is transmitted to the PEEC control module and other devices requiring vehicle speed. The buffer circuit is located close to the speed pickup to minimize noise interference, figure 11.

PEEC Vehicle Speed Buffer

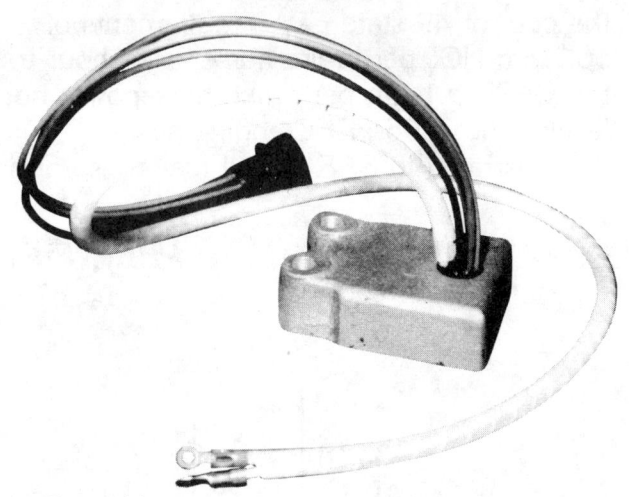

Figure 11

Exhaust Brake Enable Switch

A special exhaust brake enable switch has been designed for use with PEEC engines. The brake enable switch prevents use of an exhaust brake when any fuel is being delivered to the engine. The switch mounts on the fuel pump housing and detects shutoff fuel rack position. The switch can be used with either a mechanical or electronically governed engine.

Performance Advantages

An approximate 4.7% increase in truck fuel consumption occurs as mechanically controlled engine emissions are reduced from the current 49 state EPA requirements of 10.7 gm NO_x per brake horsepower hour to the CARB 5.1 gm per brake horsepower hour levels. The 3406 PEEC engine has demonstrated in SAE type II fuel consumption tests that the fuel penalty of stringent emissions can be significantly reduced. Adding the benefits of road speed limiting to the basic electronic engine control strategies results in a 5-12% fuel savings over a mechanical controlled engine at the CARB emission levels, figure 12.

PEEC Fuel Consumption Benefits
3406 NO_x vs Fuel Mileage

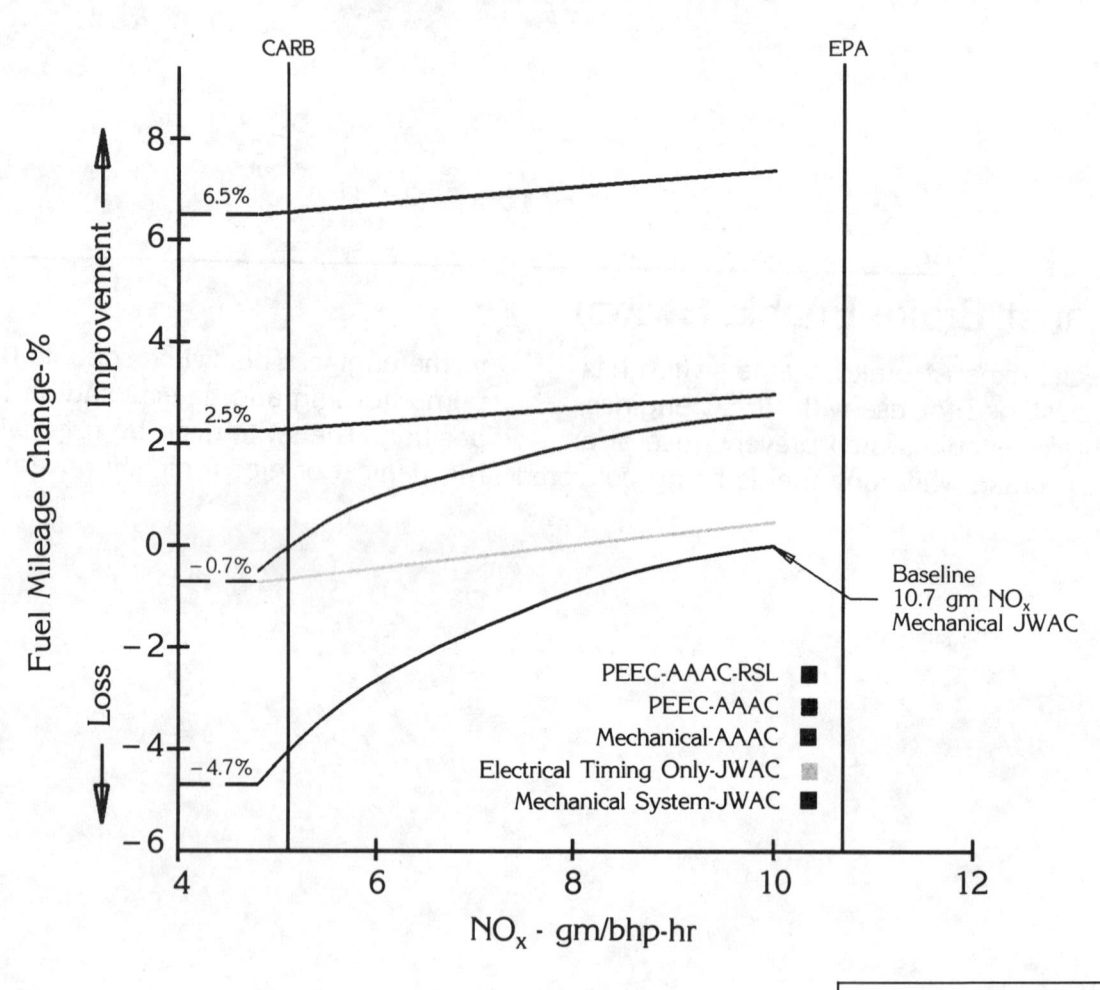

JWAC-Jacket Water Aftercooling
AAAC-Air to Air Aftercooling
RSL-Road Speed Limiting

Figure 12

Various vehicle performance test procedures have been developed in the past to measure engine response and driveability (2). The results of four such tests are shown in figure 13. Sophisticated timing and fuel air ratio control strategies are employed by PEEC to achieve an average 10% improvement in engine response. Mechanical fuel air ratio controls may restrict engine performance during part of the engine operating cycle because they must be set to control the highest emission condition. The PEEC system provides the flexibility to limit emissions under the worst conditions yet not restrict engine performance when high emissions are not produced.

Truck noise levels using SAE J366-B Pass-by Test Procedure can be lowered by controlling engine acceleration rates. The PEEC system provides the ability to compromise driveability for sound level reductions. Significant truck pass-by sound level reductions have been achieved without seriously affecting driveability.

Isochronous governing at low idle provides full engine torque capability for improved load startability while on a steep grade. Improved stability and better low idle lug than the mechanical governor allows low idle engine speed for the 3406 to be decreased from 700 to 600 rpm reducing fuel consumption. Isochronous governing is also used at rated speed to reduce fuel consumption and noise by eliminating engine operation in the mechanical overrun region. The PEEC governor operates like the Cat mechanical governor at engine speeds between low and high idle to provide good driver feel.

PEEC Driveability Improvement
3406 PEEC
70220 lbs GVW

Test	Description	% Improvement PEEC vs Mechanical
Throttle delay	Time to accelerate from low idle to rated speed while in gear	1st gear 9.4 2nd gear 9.4 3rd gear 11.6 4th gear 11.6 5th gear 9.1
Throttle interrupt	Steady state, direct gear, close throttle for 5 seconds, time to 85% rated rack	14.4
Boost evaluation	Steady state, direct gear, close throttle for 5 seconds, time to 85% rated boost pressure	9.0
Hill upshift	Time from dead stop to cover 1 mile climbing 5% grade	5.6

Figure 13

Performance Advantages

The driver comfort feature of cruise control has been incorporated in the PEEC system. Cruise control reduces driver fatigue and improves vehicle ride by eliminating throttle bounce. Two versions of cruise control will be available. The standard isochronous vehicle speed governor controls vehicle speed to within ± 1 mile per hour of the set speed within the engine's power capability. The economy cruise control strategy allows larger engine speed variations to occur thus reducing demand horsepower. Both cruise control strategies can reduce fuel consumption and driver to driver variations in a large fleet.

Road speed limiting is a significant factor in reducing vehicle fuel consumption. Road speed limiting allows a fuel efficient, gear-fast-run-slow truck specification to be used effectively when driver control programs cannot be utilized. Road speed limiting can limit top vehicle speed effectively while still specifying truck drive trains to operate the 3406 in its most efficient speed range of 1200-1400 rpm.

The PEEC system includes a data link intended for communication with other microprocessor based devices that are compatible with the proposed American Trucking Association and SAE standard. The data link can reduce duplication of truck sensors by allowing controls to share information. Information that will be available from the PEEC system on the data link includes: engine speed, vehicle speed, turbocharger inlet pressure, manifold boost pressure, rack position, throttle position, oil pressure, status word, diagnostic word, engine identification and customer specified parameters.

Serviceability

The PEEC system has been designed to have reliability equal to or better than today's mechanical systems. The electronic control system improves overall engine reliability by eliminating the troublesome mechanical components such as the throttle linkage between the cab and the engine. The PEEC system has been designed to minimize tampering. Reduced vehicle and/or engine speeds result when inputs to the PEEC system are out of specification. The engine protection features reduce engine speed and power output when low oil pressure or inlet air pressure conditions occur. The PEEC system does not shut the engine down if engine speed can be controlled. Limp home capability is provided to insure that the vehicle can be moved out of a potentially dangerous situation.

The PEEC system provides diagnostic information with or without sophisticated service tools. The control module is programmed to run diagnostic tests on all inputs and outputs to partition a fault to a specific circuit (example, fuel shutoff solenoid or the harness connecting it to the control module). Once a fault is detected, special connectors or jumper wires in the data link harness can be used to provide a flashing coded display on the dash mounted diagnostic lamp. A multimeter can be used to check the harness back to the fault and repair the problem. The rack and timing sensors can also be calibrated using the diagnostic lamp.

To facilitate faster servicing of the PEEC system, an inexpensive hand held tool is being developed to provide a digital readout of the fault mode. The tool plugs directly into the data link to provide the information to the service technician under hood. The hand held service tool also provides a direct readout of sensor calibration information.

To program the non-volatile memory a PEEC programming tool has been developed. The tool plugs into the data link to communicate with the control module microprocessor. The PEEC programming tools performs the same service functions as the hand held tool. The programming tool can also be set to display real time values of all information available on the data link for diagnosing engine problems, figure 14.

Caterpillar is also developing a universal service tool which will provide complete system analysis and support. The universal service tool can be used to simulate the actuators and sensors to test the control module or to simulate the control module to test the actuators and sensors. Both the Universal service tool and PEEC programmer will be capable of ride-along diagnostics for detecting intermittent faults, figure 15.

PEEC Programming Service Tools

Figure 14

Caterpillar Universal Electronic Control Service Tool

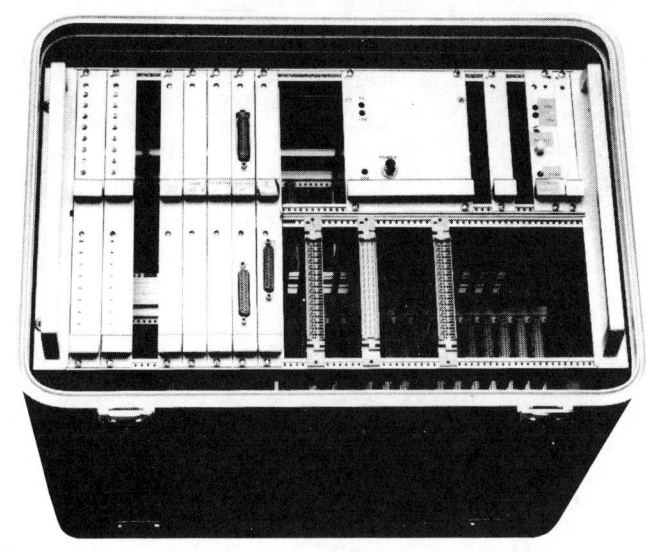

Figure 15

Availability

The PEEC system will be introduced on 3406 CARB engines for the 1987 truck model year with ratings from 285 to 425 hp (213 to 317 kW). All 3406 Truck Engine ratings will be available with PEEC by the 1988 truck model year. PEEC engines are being field tested now to insure that the design meets the reliability and customer expectation goals.

Summary

The 3406 PEEC engine has demonstrated the ability to meet our emissions performance fuel consumption and reliability development goals. PEEC provides a good return on investment to the customer by reducing fuel consumption, reducing operating cost, providing easy maintenance and good reliability. PEEC provides information and features not previously available on diesel engines. PEEC is the first of many progressive steps for electronic engine controls on fuel efficient diesel engines of the future.

References:
1) D.H. Connor and R.A. Stapf, "Caterpillar 3406B Truck Engine", SAE Paper 831201 (1983).
2) M.C. Brands, J.R. Werner, J.L. Hoehne and S. Kramer "Vehicle Testing of Cummins Turbocompound Diesel Engine", SAE Paper 810073 (1981).

3406 PEEC Engine

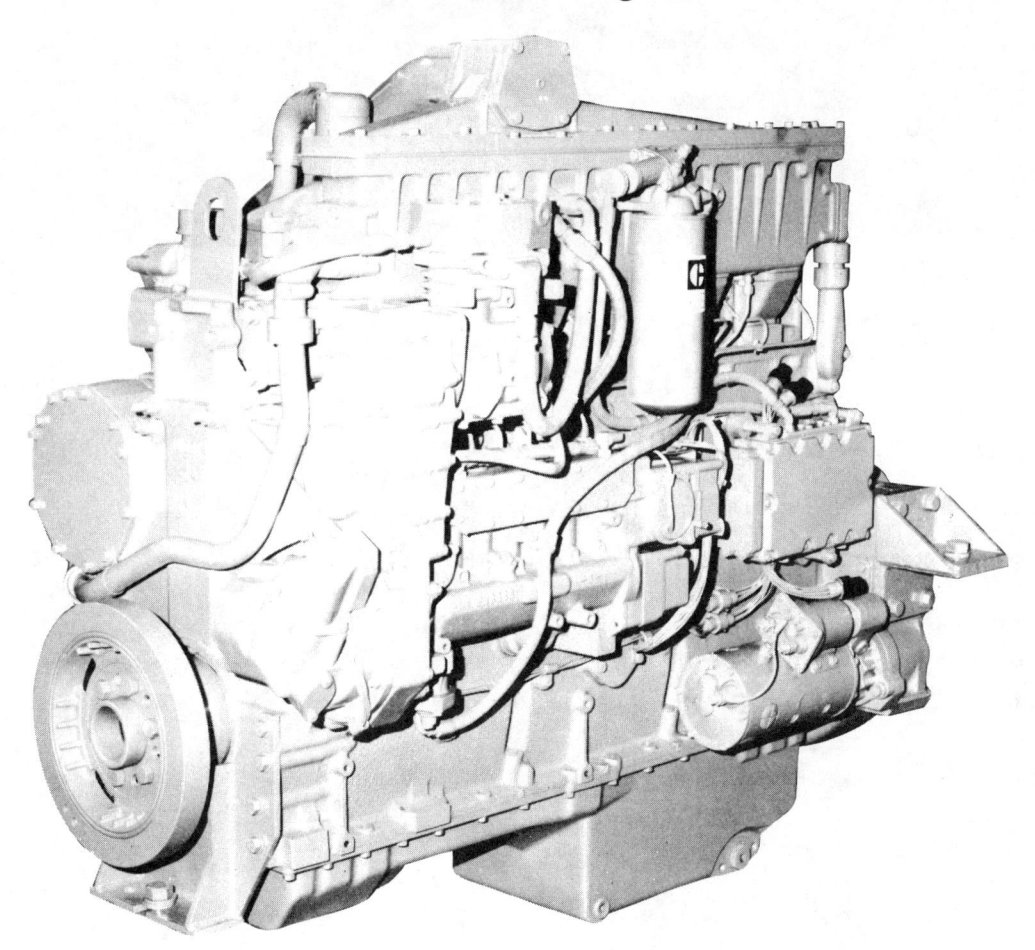

850174

A Digital Control Algorithm for Diesel Engine Governing

Daniel C. Garvey
Woodward Governor Co.
Engine and Turbine Controls Div.
Fort Collins, CO

ABSTRACT

The performance of a microprocessor based precision engine speed control system was investigated. A sample rate selection criteria is presented along with a procedure to implement a high performance digital PID control algorithm. The algorithm requires a digital speed sensor of 12 to 14 bits to minimize excessive fuel rack motion at steady state due to digital quantization effects. Computer simulation and experimental test results of the algorithm are presented for an 1800 RPM, 125 Kilowatt engine generator set.

INTRODUCTION

The use of microprocessors (μP) in industrial control systems has become very popular for several reasons. A μP can reduce the number of circuit components and cost in a design and increases system reliability and control system design flexibility. When a μP based control system is applied to an industrial diesel engine propulsion system or electric power generation system, several performance requirements must be satisfied. The control system transient response must be rapid to minimize engine speed deviations due to sudden changes in load. During steady state operation, the control system must maintain precise control of speed with a minimum of fuel rack motion. These requirements must be satisfied in the presence of engine torsional oscillations in speed which can be severe.

These high standards impose constraints upon the design requirements of a μP based control system. A digital control system which meets these standards requires the selection of

— a μP sample rate
— a control algorithm
— a precision speed sensor
— a μP computation method
— a particular μP

This paper presents a sample rate selection criteria and describes the design of a μP based PID control algorithm which yields rapid transient response and satisfies the steady state requirements of precision speed control with a minimum fuel rack motion. The requirements of the speed sensor necessary to implement the algorithm are identified in terms of an operating range and number of bits. Computer simulation and experimental engine test results obtained with the digital controller are presented.

ANALYSIS

A block diagram descriptive of the engine control system is shown in Figure 1. To aid in the design and evaluation of different digital control algorithms, a sample rate selection criteria was determined and a mathematical model descriptive of the system response was developed. Analytical expressions describing each of the blocks shown in Figure 1 are developed in the mathematical model. Mathematical symbols are identified in the Nomenclature tabulation:

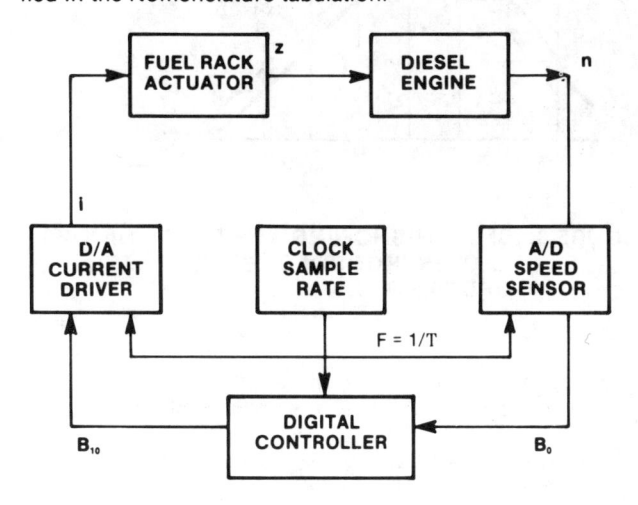

FIGURE 1 DIESEL ENGINE DIGITAL SPEED CONTROL SYSTEM.

MICROPROCESSOR SAMPLE RATE SELECTION — A measure of the performance of an analog control system is given by inspection of the open loop transfer function

$$G(s) \cdot H(s) \qquad (1)$$

to obtain measures of crossover frequency (F_O) and phase margin (θ). In a high performance regulator loop, a phase margin of

$$\theta = 30 \text{ (degrees)}$$

is desirable at a high open loop crossover frequency (F_O). Values of phase margin much less than 30 degrees are undesireable and cause stability problems in the control system. A 0 degree phase margin demonstrates instability.

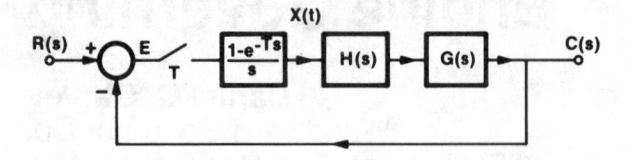

FIGURE 2. SAMPLING IN AN ANALOG CONTROL

The effect of sampling on this system is evaluated by considering an equivalent digital system shown in Figure 2. (1)* Time domain plots of a signal and its sampled and clamped resultant are shown in Figure 3. The boxcar output of the zero order hold follows the input curve delayed by T/2 seconds, with T the sample period. The open loop transfer function with sampling is given by

$$G(s) \cdot H(s) \, e^{-(T/2)s} \qquad (2)$$

where

$$T \triangleq \text{sample time}$$

$$F = 1/T \triangleq \text{sample frequency}$$

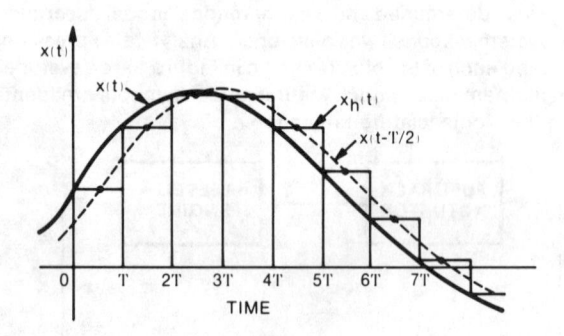

FIGURE 3. SKETCH SHOWING THE DELAY OF T/2 SEC INTRODUCED BY THE ZERO-ORDER HOLD.

In an open loop Bode plot, the additional phase shift at crossover frequency (F_O) due to sampling is given by

$$\Delta\theta = F_O T \, 180 \quad \text{(degree)} \qquad (3)$$

This phase shift is a reduction in phase margin at the crossover frequency F_O. If the open loop crossover frequency is reduced by ΔF to improve the phase margin in the presence of sampling, then

$$F_O = F^* \left(1 - \frac{\Delta F}{F^*}\right) \qquad (4)$$

---

*Numbers in parentheses designate references at the end of the paper.

Let

$$a = \frac{\Delta F}{F^*}$$

the phase shift at crossover frequency is given by

$$\Delta\theta = \frac{180 \, F^* \, (1-a)}{F} \quad \text{(Degree)} \qquad (5)$$

The ratio of sample frequency to open loop crossover frequency is given by

$$\frac{F}{F^*} = \frac{180 \, (1-a)}{\Delta\theta} \qquad (6)$$

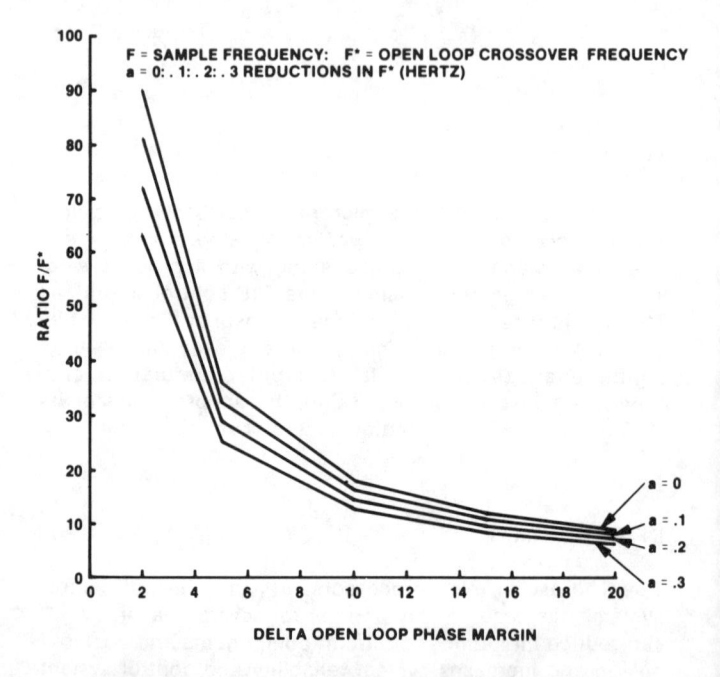

FIGURE 4
RATIO (F/F*) AS A FUNCTION OF PHASE MARGIN LOSS AND REDUCTION IN CROSSOVER FREQUENCY (1-a).

This equation is plotted in Figure 4 as a function of

— the reduction in phase margin $\Delta\theta$

— the reduction in the open loop crossover frequency (1-a)

A small high speed diesel engine is characterized by an open loop crossover frequency of

$$F^* = 4.0 \text{ Hertz}$$

Allowing a minor phase margin reduction of

$$\Delta\theta = 5°$$

and no reduction in bandwidth, a = 0

$$\frac{F}{F^*} = 36$$

and

$$F = 144 \text{ Hertz Sample Frequency}$$

or

$$T = 1/F = .0069 \text{ seconds}$$

the required µP sample time for a small high speed diesel engine. If a sample time is selected for a small high speed engine, T = .016 second then

$$\frac{F}{F^*} = 15.625$$

and from Figure 4, the phase margin is reduced by 12° at a = 0. This is a major reduction in phase margin and yields a ringing type of response.

In a large bore, low speed marine propulsion engine

$$0.2 \leq F^* \leq 0.6 \text{ Hertz}$$

and is a function of the engine speed range. With F/F* = 36 then

$$21.74 \geq F \geq 7.25 \text{ Hertz}$$

for a µP sample rate on a low speed marine engine.

These results demonstrate that in a high performance digital control system it is desireable to satisfy the design criteria

$$15 \leq \frac{F}{F^*} \frac{\text{(Sample Frequency)}}{\text{(Open Loop Crossover Frequency)}} \leq 36 \quad (7)$$

to minimize control system loss in phase margin or reduction in crossover frequency. Figure 5 depicts a plot of the required digital control sampling frequency (F) versus control loop crossover frequency (F*) satisfying the above constraint.

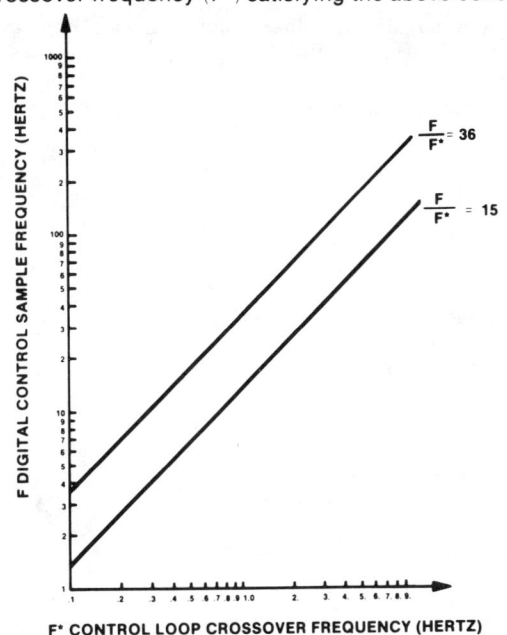

FIGURE 5. MICRO PROCESSOR SAMPLE FREQUENCY (F) AS A FUNCTION OF OPEN LOOP CROSSOVER FREQUENCY (F*).

DIESEL ENGINE — The steady state performance of a typical diesel engine is shown in Figure 6. In this figure, engine power is plotted versus engine speed at several different positions of the fuel rack. The engine speed governor forces operation along the dashed lines for either droop or isochronous speed control. Droop operation is required during parallel operation of two mechanical drive units or with the commercial utility. Speed droop requirements of 5 to 10 percent establish the minimum range necessary for the governor speed sensor. The dynamic response of a diesel engine is described by a normalized torque balance on the engine flywheel.

$$t - L = \frac{sn}{\Upsilon} \quad (8)$$

with engine acceleration coefficient

$$\Upsilon = \frac{30 L_E}{\pi J N_R}$$

The engine developed torque is proportional to fuel rack position but is delayed by engine dead time

$$t = z \, e^{-\tau_1 s} \quad (9)$$

The engine dead time is characterized by

$$\frac{15}{N_R} \leq \tau_1 \leq \frac{15}{N_R} + \frac{60}{N_R Q}$$

where

N_R = engine RPM
Q = number of cylinders firing per revolution

The position of the fuel rack is proportional to the applied current but delayed by the electrohydraulic actuator time constant (τ_2)

$$z = \frac{i}{\tau_2 s + 1} \quad (10)$$

Representative values of engine and fuel system constants for a small high speed engine are given by $\Upsilon = .51$, $\tau_1 = .015$ and $\tau_2 = .08$.

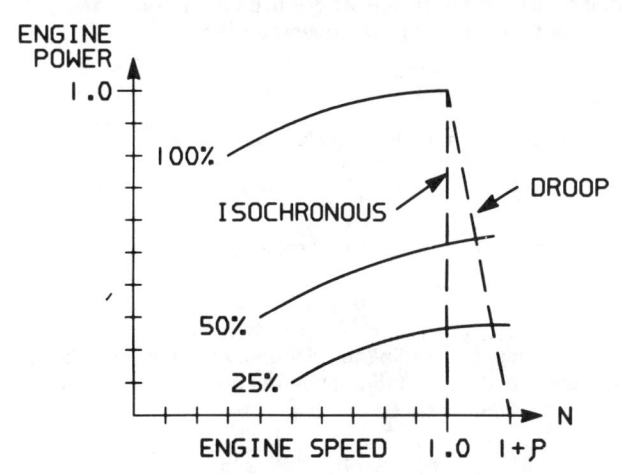

FIGURE 6: DIESEL ENGINE PERFORMANCE MAP WITH SPEED GOVERNOR.

SPEED SENSOR — A magnetic pickup or proximity sensor is commonly used to obtain a measure of engine speed. The magnetic pickup output is a sine wave at a frequency proportional to engine speed. This frequency is distorted by engine torsional oscillations at crankshaft speed and firing frequency. A frequency to voltage converter with an analog to digital convertor provides a digital measurement of speed. The measurement accuracy of a digital speed sensor is proportional to the operating speed range of the engine and the digital word length. The output of an ideal digital speed sensor is given by

$$B = \text{Integer}\left[\left(\frac{n-n_1}{n_2-n_1}\right) \cdot 2^N\right] \quad (11)$$

where

n = actual speed per unit
n_1 = minimum speed per unit
n_2 = maximum speed per unit
and N = digital word length

The per unit gain of the digital speed sensor is given by

$$\frac{\Delta B}{2^N} \cdot \frac{1}{\Delta n} = K_D \quad (12)$$

with

$$K_D = \frac{1}{n_2-n_1} \quad (13)$$

The minimum per unit gain of the digital speed sensor can be defined in terms of the incremental measurement accuracy at steady state. For an incremental accuracy requirement of

$$\Delta n = .0003$$

and a 12 bit speed sensor,

$$(K_D)_{min} = .813$$

With a 14 bit speed sensor and the same incremental measurement standard

$$(K_D)_{min} = .203$$

Similarly, a maximum value of per unit gain in the speed sensor can be defined in terms of an operating range

$$n_2 - n_1$$

The minimum acceptable range is given by

$$n_2 - n_1 = .1$$

such that for either a 12 or 14 bit speed sensor

$$(K_D)_{max} = 10.0$$

The requirements of a minimum accuracy measurement and a minimum speed range define a per unit gain range for either a 12 or a 14 bit digital speed sensor

$$.813 \leq K_D \leq 10. \quad \text{for 12 bits}$$

and

$$.203 \leq K_D \leq 10. \quad \text{for 14 bits}$$

The measured value of the speed must be scaled to the accuracy of the computational algorithm

$$B_O = 2^{M-N} \cdot B \quad (14)$$

CONTROL ALGORITHM SELECTION - The μP control algorithm selected follows the format of an analog PI or PID controller. This algorithm was selected because of its familiarity, applicability to the majority of industrial control problems and proven results. Several different forms of the PID control algorithm exist. For tuning and analytical purposes, a non-interacting form of the PID algorithm in the frequency domain is desirable. (2) The PID control algorithm selected is described by a speed sensor.

$$V = K \cdot n \quad (15)$$

and the control

$$i = \frac{I}{s} \cdot \frac{(\beta s + 1) \cdot (\delta s + 1)}{(ps + 1)} \cdot \left[V_R - \frac{V}{ps + 1}\right] \quad (16)$$

The product $(I \cdot K)$ represents per unit gain on a Bode diagram, $1/\delta$ and $1/\beta$ are adjustable zero locations and $1/p$ characterizes the roll off filter pole location.

An equivalent digital controller algorithm can be obtained by using one of several different methods to convert a continuous transfer function to its discrete form. (3) A simple method is given by the backward difference approach, which substitutes

$$s = \frac{1-z^{-1}}{T}$$

in the analog transfer function expression. This substitution is applicable to systems which satisfy Shannons Sampling Theorem or the Nyguist criteria

$$p \geq p_{nyquist} = \frac{T}{\pi}$$

With the above substitution, the analog PID algorithm was converted to a digital PID form.

The measured speed signal applied to a digital filter

$$B_1 = B_1 z^{-1} + \left(\frac{T}{p+T}\right) \cdot \left(B_O - B_1 z^{-1}\right) \quad (17)$$

The digital measure of error

$$B_2 = B_{REF} - B_1 \quad (18)$$

The lead and lag or integration given by

$$B_4 = B_2 + (\beta/T) \cdot (B_2 - B_2 z^{-1}) \quad (19)$$

$$B_6 = B_6 z^{-1} + (IT) \cdot B_4 \quad (20)$$

The lead and lag combination

$$B_8 = B_6 + (\delta/T) \cdot (B_6 - B_6 z^{-1}) \quad (21)$$

$$B_9 = B_9 z^{-1} + \left(\frac{T}{p+T}\right) \cdot \left(B_9 - B_9 z^{-1}\right) \quad (22)$$

CURRENT DRIVER - The performance of the digital to analog convertor and current driver is given by

$$B_{12} = \text{Integer}\left(B_9 \frac{2^R}{2^M}\right) \quad (23)$$

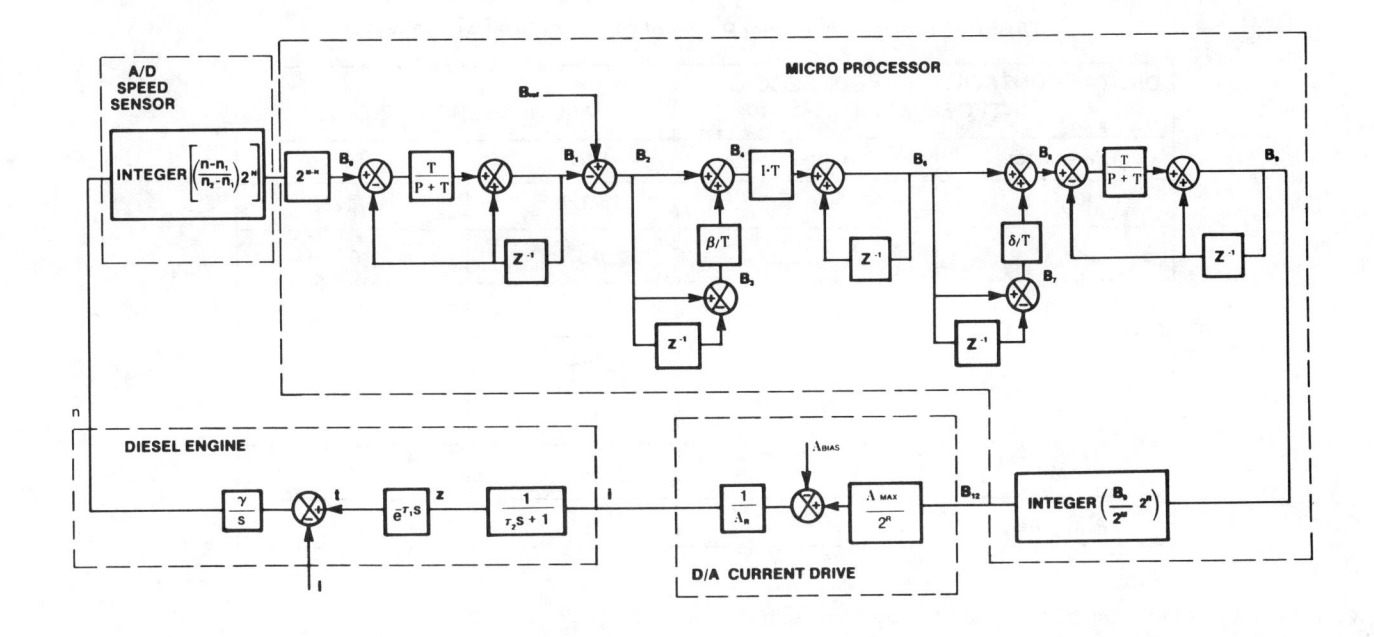

FIGURE 7. MATHEMATICAL MODEL OF A DIESEL ENGINE DIGITAL CONTROL SYSTEM

$$A = \left(\frac{B_{12}}{2^R}\right) \cdot A_{max} - A_{Bias} \quad (24)$$

and when normalized

$$i = \frac{A}{A_R} \quad (25)$$

The above system of equations presents a mathamatical model which can be used in the evaluation of digital control system concepts. A mathematical block diagram depicting this model is shown in Figure 7.

Typical values of adjustment range for the analog control and the equivalent digital control with a sample time of 8 milliseconds are shown in Tables 1 and 2. With an 8 millisecond sample rate, the time required for the microprocessor to execute a control algorithm must be minimized. A minimum algorithm execution time along with a wide computational range is achieved with 16 bit fixed point mathematics. To avoid overflow, computations were performed with a 24 bit word. The adjustment ranges and typical settings for a small high speed engine are shown in Table 2.

DIGITAL CONTROL SYSTEM RESPONSE - The quantization effects of

- input speed sampling (N = 12 or 14 bits)
- output current sampling (R = 12 bits)
- fixed point mathematics (M = 16 bits)

Lead to errors in the control system which cause low frequency (.1 to 4 Hertz) limit cycle oscillations in fuel rack motion and engine speed.

Large amplitude low frequency fuel rack oscillations are objectionable due to excessive linkage or fuel pump wear and the possibility of exciting power transmission system natural frequencies. A limit cycle amplitude of less than .005 per unit fuel rack motion is desirable. To achieve the design goal of .005 per unit fuel rack motion with a 1 bit error in measurement accuracy at the speed sensor, it is necessary to properly distribute total digital controller gain ($K_D \cdot I \cdot T$) between the digital speed sensor (K_D) and the digital integrator (IT) for both the PI and the PID algorithms. The design approach implemented satisfies the total controller gain

$$(K\ I)_A = K_D \cdot I$$

TABLE 1: Typical Adjustment Range of Analog Control Algorithms

| ANALOG CONTROL || SMALL HIGH SPEED ENGINE ||
Parameter	Range	PID	PI
$(K\ I)_A$	2.0 to 500.	60.	16.
δ	.008 to 0.2	.128	0.0
β	.008 to 2.0	.35	.512

TABLE 2: Typical Adjustment Range of Digital Control Algorithms

DIGITAL CONTROL and T = .008 SECOND INTEGER MATH		SMALL HIGH SPEED ENGINE	
Parameter	Range	PID	PI
$K_D I \cdot T$	$1/2^6$ to 2^2	60/128	16/128
δ/T	1 to 32	16	0
β/T	1 to 256	44	64
P		.008	.008
R		12	12
M		16	16
(A_{max}/A_R)		2	2

defined by the algorithm by first computing the digital integrator gain I which meets the minimum specified fuel rack motion requirement when the speed sensor error is one bit. The per unit gain of the digital speed sensor is then computed as

$$K_D = \frac{(KI)_A}{I}$$

and must lie within the constraints

$$(K_D)_{min} \leq K_D \leq (K_D)_{max}$$

The digital algorithm deviation equations for a 1 bit change in measured speed are obtained from Figure 7. The computational control algorithm yields for a 1 bit error

$$\Delta B_g = 2^{M-N} \left(\frac{T}{P+T}\right)^2 (1 + \delta/T) \cdot (1 + \beta/T) \cdot (IT) \cdot 1 \quad (26)$$

and the current driver is given by

$$\Delta i = \frac{A_{max}}{A_R} \cdot \left(\frac{1}{2^R}\right) \text{Integer} \left(\frac{\Delta B_g}{2^{M-R}}\right) \quad (27)$$

Substituting representative values from Table 2, into Equation (27), the constraint on fuel rack motion

$$\Delta i = .005$$

is satisfied with

$$\Delta B_g = 160$$

The computational control algorithm requires that the gain on the digital integrator satisfy

$$I = \frac{\Delta B_g}{T \left[2^{M-N} \left(\frac{T}{p+T}\right)^2 (1+\delta/T)(1+\beta/T) \right]} \quad (28)$$

Solving the above equation with values from Table 2, for the PID algorithm

$$(I)_{PID} = 6. \text{ at } N = 12$$

and

$$(I)_{PID} = 26. \text{ at } N = 14$$

For the PI algorithm

$$(I)_{PI} = 76 \text{ at } N = 12$$

and

$$(I)_{PI} = 307 \text{ at } N = 14$$

The gain of the digital speed sensor for the PID algorithms is then given by

$$K_D = \frac{(KI)_A}{(I)_{PID}}$$

$$K_D = 10.0 \text{ at } N = 12$$

and

$$K_D = 2.3 \text{ at } N = 14$$

Both of these values lie within the allowable range of the speed sensor per unit gain but imply a narrow operating speed range.

If the minimum value of per unit gain for a 12 bit speed sensor had been selected to maximize the speed range, then

$$I = \frac{60}{.813}$$

$$I = 73.$$

With this value in the PID algorithm, the motion of the fuel rack due to a 1 bit error obtained from equations (26) and (27) is given by

$$\Delta B_g = 1745.$$

and

$$\Delta i = .053$$

This amplitude of fuel rack motion is unacceptable.

TABLE 3: Digital PID Control System Parameters for a Small High Speed Engine

Parameter	Digital PID Algorithm T = .008 Second Speed Sensor		Digital PID Algorithm T = .016 Second Speed Sensor	
	N = 12 Bit	N = 14 Bit	N = 12 Bit	N = 14 Bit
ΔB	1	1	1	1
ΔB_g	160	160	160	160
Δi	.005	.005	.005	.005
B_{ref}	32768	42672	32768	53692
$n_2 - n_1$	.1	.43	.2	.83
Δn	2.4×10^{-5}	2.6×10^{-5}	4.0×10^{-5}	5.0×10^{-5}
I	6.	26.	12.	50
K_D	10	2.3	5.	1.2
β/T	44	44	18	18
δ/T	16	16	8	8

To increase the operational speed range of the PID algorithm, the μP sample time can be increased to T = .016 second with some reduction in system transient response. This allows the digital integrator gain to be increased for both the 12 and 14 bit speed sensors,

I = 12   at 12 bits

and

I = 50   at 14 bits

The per unit gain for each speed sensor is given by

K_D = 5.0   at 12 bits

and

K_D = 1.2   at 14 bits

Table 3 presents a summary of the digital PID algorithm parameters for two different speed sensors and two different sample periods. As the number of bits in the digital speed sensor increases or as the sample period T increases in the PID algorithm, the gain of the digital integrator increases and minimizes the per unit gain of the digital speed sensor.

Employing a similar analysis for the PI algorithm, the gain of the digital speed sensor is given by

$$K_D = \frac{(K_A \cdot I_A)}{(I)_{PI}}$$

K_D = .21   at N = 12

and

K_D = .05   at N = 14

Both of these values are less than the minimum allowable per unit values of speed sensor gain. This result demonstrates that the measurement accuracy requirement on speed cannot be met. Selecting the minimum allowable speed sensor gain to satisfy the accuracy requirement for both the 12 and 14 bit sensors

K_D = .813

The PI algorithm then requires

$$I = \frac{(K_A I_A)}{K_D}$$

I = 20

The resulting fuel rack motion due to a 1 bit error is given by Equations (26) and (27),

at 12 bits

ΔB_g = 40

Δi = .0015

and at 14 bits

ΔB_g = 10

Δi = .0005

A summary of the digital PI algorithm parameters for two different speed sensors is presented in Table 4.

TABLE 4: Digital PI Control System Parameters for a Small High Speed Engine

Parameter	Digital PI Algorithm T = .008 Second Speed Sensor	
	N = 12 Bit	N = 14 Bit
ΔB	1	1
ΔB_9	40	10
Δi	.0015	.0005
B_{ref}	53280	53280
n_2-n_1	1.23	1.23
Δn	$3. \times 10^{-4}$	7.5×10^{-5}
I	20	20
K_D	.813	.813
δ/T	64	64
δ/T	0	0

COMPUTER SIMULATION STUDY — A computer simulation of the mathematical model depicted in Figure 7 was performed to evaluate both the transient and the steady state performance of the two control algorithms. The simulation subroutine describing the PID algorithm is described in Appendix 1. Figure 8 presents simulation results for a start to rated speed and addition of 100% load, with the PID control algorithm, a 12 bit speed sensor and T = .008 seconds. The transient response of per unit engine speed (n) and actuator drive current (i) are depicted. The maximum speed deviation for the PID algorithm is 3.5 percent. Figure 9 shows the steady state performance at rated speed and no load of engine speed and actuator drive current. The limit cycle oscillation at steady state is characterized by

$$|n| = 2.5 \times 10^{-5}$$
$$|i| = .005$$

at 1 Hertz. Figures 10 to 12 show similar transient results with different speed sensors and sample times. When the control algorithm is changed to PI with a 12 bit speed sensor and T = .008 seconds, the transient response of Figure 13 is obtained. The results are summarized in Table 5 below.

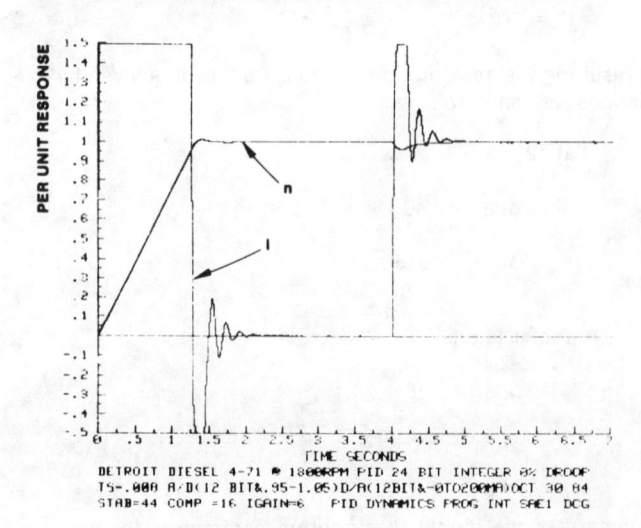

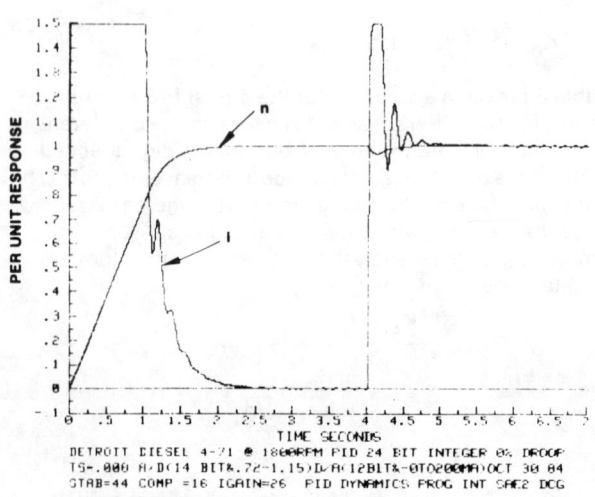

FIGURE 8 TRANSIENT RESPONSE OF A DIGITAL PID CONTROL WITH A 12 BIT SPEED SENSOR AND T=.008 SECOND.

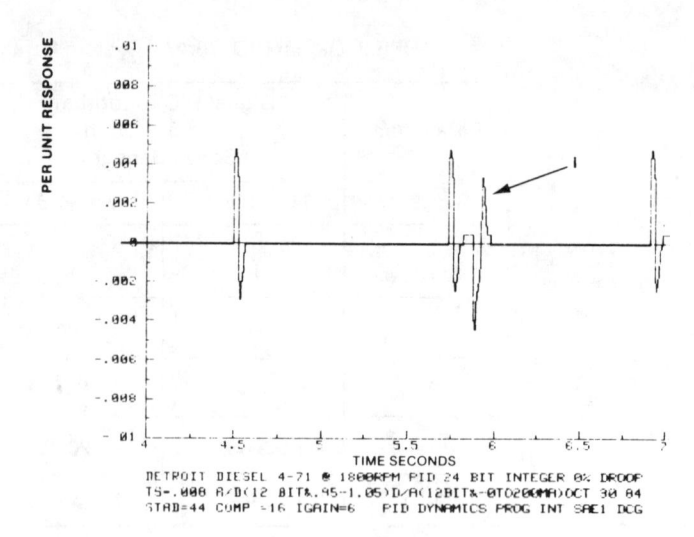

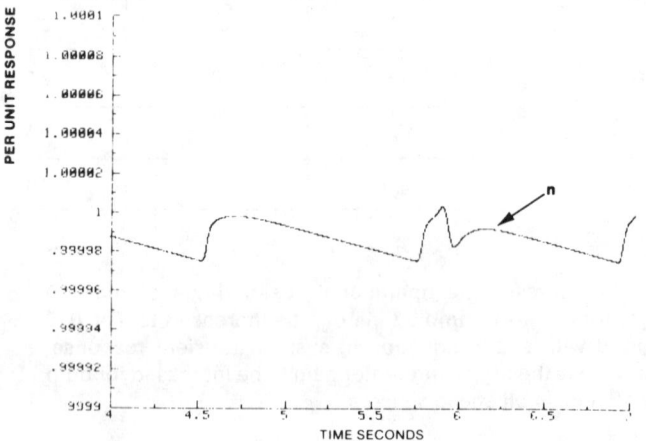

FIGURE 9 STEADY STATE RESPONSES AT A DIGITAL PID CONTROL WITH A 12 BIT SPEED SENSOR AND T=.008 SECOND.

FIGURE 10 TRANSIENT RESPONSE OF A DIGITAL PID CONTROL WITH A 14 BIT SPEED SENSOR AND T=.008 SECOND.

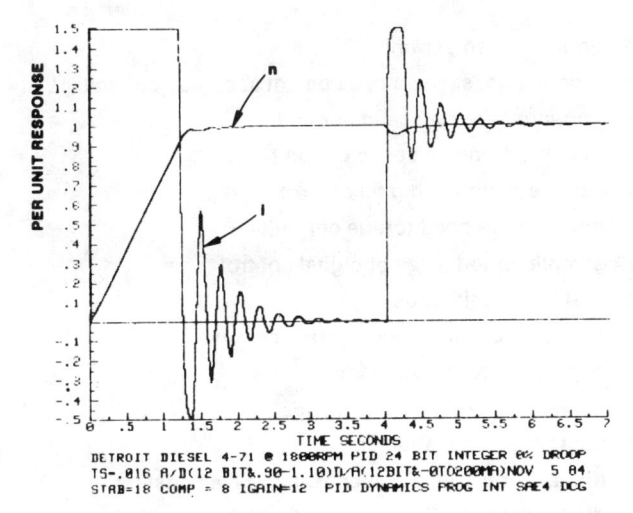

FIGURE 11 TRANSIENT RESPONSE OF A DIGITAL PID CONTROL WITH A 12 BIT SPEED SENSOR AND T=.016 SECOND.

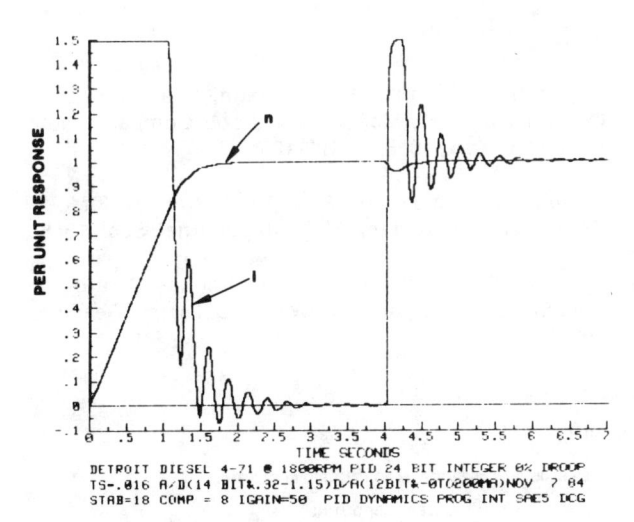

FIGURE 12 TRANSIENT RESPONSE OF A DIGITAL PID CONTROL WITH A 14 BIT SPEED SENSOR AND T=.016 SECOND.

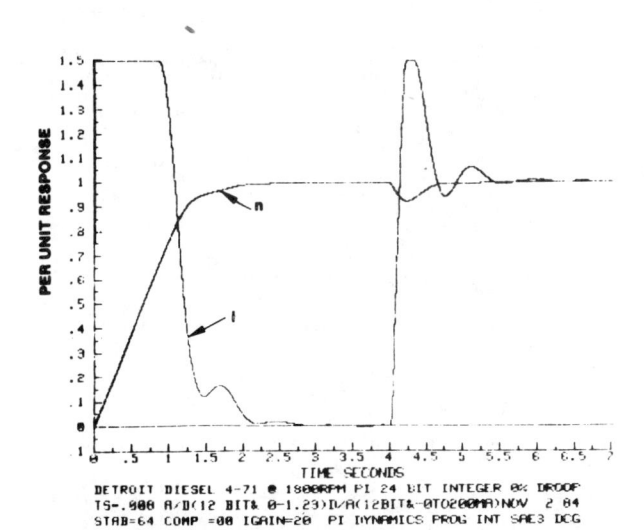

FIGURE 13 TRANSIENT RESPONSE OF A DIGITAL PI CONTROL WITH A 12 BIT SPEED SENSOR AND T=.008 SECOND.

In each case the limit cycle oscillation at steady state satisfied the design criteria $\Delta i \leq .005$.

TABLE 5: Digital Control Transient Response

ALGORITHM	SPEED SENSOR N	SAMPLE TIME T	PEAK SPEED DEVIATION ΔN	SPEED RANGE $N_2 - N_1$
PID	12	.008	.035	.10
PID	14	.008	.035	.43
PID	12	.016	.044	.20
PID	14	.016	.044	.83
PI	12	.008	.076	1.23

ENGINE TEST RESULTS — A real time simulation of the digital diesel governor has been developed using an HP 9836 computer and a HP 8940B multiprogrammer for A/D and D/A conversions. The real time model allows a rapid test of different control algorithms and concepts before hardware is designed. Control algorithms similar to Appendix 1 were implemented in PASCAL and tested with an electrohydraulic fuel rack actuator on a Detroit Diesel 4-71 engine generator set. The engine is rated at 1800 RPM and 125 Kw.

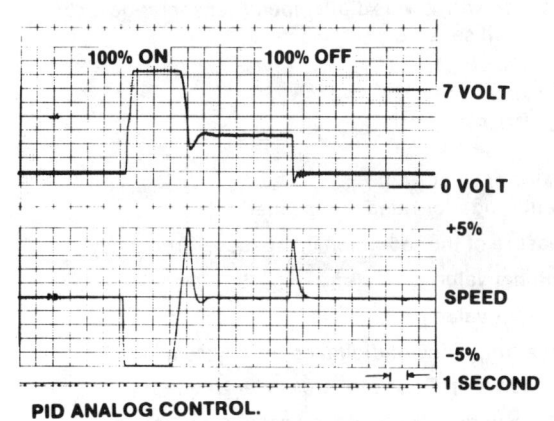

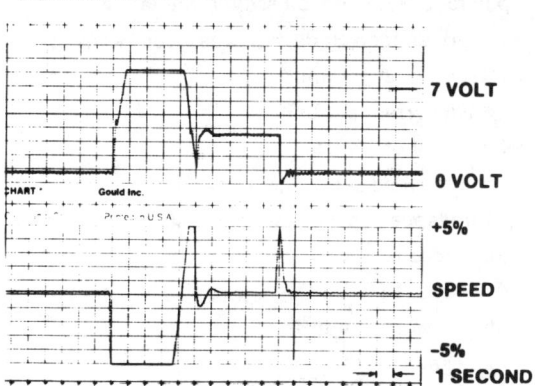

PID ANALOG CONTROL.

DIGITAL PID CONTROL WITH 12 BIT A/D SPEED SENSOR AND 8 MILLISECOND SAMPLE RATE.

FIGURE 14 EXPERIMENTAL ENGINE TEST DATA.

A simple digital speed governor was developed using an 8 bit μP, an 8 millisecond sample time and a 12 bit digital speed sensor. Figure 14 depicts a comparison of the transient and steady state response between an analog PID control and the digital control on a diesel engine. On each chart the upper trace depicts voltage applied to an actuator which positions the fuel rack and the lower trace measures engine speed. The digital governor was set with I = 4, $\beta/T = 25$, $\delta/T = 16$. The maximum speed deviations were 4.2 percent for the analog control and 4.6 percent for the digital control. The transient response and the steady state response of the two controls compare favorably.

CONCLUSIONS — A μP sample rate selection criteria has been presented and the theoretical performance of two digital control algorithms for diesel engine governing has been described. The effect of a 12 bit and a 14 bit digital speed sensor upon the operating range and steady state performance of each algorithm has been shown. A high performance digital PID algorithm provides rapid transient response with minimum fuel rack motion at steady state but imposes constraints on the digital speed sensor operational range. The speed sensor range can be increased by increasing the number of bits (N) in the speed sensor and the sample period T. For a small high speed engine, minimum PID algorithm requirements are a 12 bit speed sensor, a .008 second sample period and 0.1 per unit speed range. A 14 bit speed sensor would allow the per unit speed range to be increased to .83 at .016 second sample period.

The digital PI algorithm does not provide the rapid transient response of the PID algorithm but does yield minimum fuel rack motion with a wide digital speed sensor range for both the 12 and 14 bit sensors.

NOMENCLATURE

A	actuator drive current
A_R	current range for rated engine load
a	a measure of the reduction in crossover frequency
B	μP integer value
B_j	μP integer value j = 0, 1, ...12
F	sample frequency of digital control system
F_O	final open loop Bode plot crossover frequency
F^*	initial open loop Bode plot crossover frequency
ΔF	reduction in open loop Bode plot crossover frequency
I	control integral gain per unit
i	actuator drive current per unit
J	flywheel moment of inertia
K_A	analog control speed sensor gain per unit
K_D	digital control speed sensor gain per unit
L_E	rated engine torque
L	engine load torque per unit
M	computational word length
N_R	rated engine speed
N	digital speed sensor word length
n	engine speed per unit
Δn	incremental speed measurement accuracy per unit
n_1	minimum engine speed per unit
n_2	maximum engine speed per unit
T	sample period of digital control
t	engine developed torque per unit
Z^{-1}	sample period delay of digital control
z	fuel rack position per unit
Υ	engine acceleration coefficient per unit
τ_1	engine dead time seconds
τ_2	actuator time constant seconds
μP	microprocessor abbreviation
δ	digital control lead compensation time constant
β	digital control lead compensation time constant

REFERENCES

1) Ragazzini, J.R. and G.F. Franklin, "Sampled-Date Control Systems", McGraw Hill Book Company, New York, New York, 1958, pp 123-125.

2) DeBolt, R.R. and B.E. Powell, "A Natural 3-Mode Controller Algorithm for DDC", ISA Journal, Sept, 1966.

3) Franklin, G.F. and J.D. Powell, "Digital Control of Dynamic Systems", Addison-Weley Publishing, Reading, MA, 1980.

APPENDIX 1

COMPUTER SIMULATION PROGRAM FOR THE PID ALGORITHM.

```
PID__cont:   ! PID CONTROLLER SUBROUTINE
             ! WORDSIZE IN COMPUTATIONS  24 BITS
        Bb=( (N__pm-.95)/(1.05-.95) )•4096    !SPEED SENSOR 12 BIT
        Bb=INT(Bb)
             IF Bb>4096 THEN Bb=4096          !CLAMP ON
             IF Bb<0 THEN Bb=0                !RANGE 12 BIT
        B0=16•Bb
        B1=B1z+( (B0-B1z)/2)                  !FILTER TIME CONST .008
        B1=INT (B1)
        B2=B__ref-B1
        B3=INT (B2)-INT (B2z)                 !STABILITY
        B4=INT (B__t•B3)+INT (B2)
        B5=INT (B4•Igain•2/256)
             IF B5>B__sat THEN B5=B__sat      !CHECK FOR OVERFLOW
             IF 5<-B__sat THEN B5=B__sat
                                              !INTEGRATION
        B6=INT (B5)+INT (B6z)                 ! INTEGRATOR OUTPUT
             IF B6>B__max THEN B6=B__max      !CLAMP ON
             IF B6<B__min THEN B6=B__min      ! INTEGRATOR
        B7=INT (B6)-INT (B6z)
        B8=INT (B6)+INT (B7•D__t)
             IF B8>B__max THEN B8=B__max      !CLAMP ON
             IF B8<B__min THEN B8=B__min      !OUTPUT
        B9=(B9z)+(B8-B9z)/2     !.008 FILTER@ OUTPUT
        B12=INT (B9/16)         !16 BIT TO 12 BIT D/A
        Ma=200•(B12/4096-50.2)  !OUTPUT CURRENT 50 MA BIAS (0,200MA)
        I__act=Ma/Ir            !NORMALIZED OUTPUT CURRENT
        SAVES
             B1z=INT (B1)    ! FILTER
             B2z=INT (B2)    !STABILITY
             B6z=INT (B6)    !!INTEGRATOR
             B9z=INT (B9)    !FILTER
        RETURN
        ! ···················· END PID DIGITAL CONTROLLER LOOP···················
```

860141

A New Concept for Electronic Diesel Engine Control

Alois Kainz, Artur Seibt, and Christian Augesky
Voest-Alpine Friedmann GmbH
Vienna, Austria

ABSTRACT

The first model of a new series of electronic diesel engine control units (EDCUs) is described:
First time incorporation of a powerful failsafe computer architecture with 2 to 4 third generation 8/16 bit microcomputers solves the safety problem inherent in diesel engine control and features EGR-control, injection timing, cruise control etc. in a small package.

INTRODUCTION

The usage of electronic diesel engine control units hinges upon the guarantee of safety and reliability of operation. In case of an electronics malfunction traffic accidents with personal injury and/or engine damage may occur. Under no-load conditions modern unit injector diesel engines may accelerate from idle speed to a speed on the borderline of damage within 0.5 seconds even with very little fuel injected.

It is assumed that the diesel engine is provided with an emergency stop mechanism which forces the engine into standstill by a command from the control unit.

LIMITS OF PRESENT CONCEPTS

Electronic diesel engine control units currently used on engine test stands or in test vehicles are based upon single microcomputer architectures consisting of one microcomputer with a number of additional peripheral units, e. g. memories, timer/counter units or A/D converters. The microcomputer control program and the fixed parameters specific to an engine and a vehicle are stored in the same read-only memory. This variety of peripheral units is connected to the microcomputer's address and data buses which run all over the electronics and are prone to pick-up of electrical interference. A single spike, however, may be sufficient to cause program and, hence, EDCU malfunction.

The so-called "watch-dog", supposed to reset the microcomputer in such an event is very slow to react, further it will not always be activated leaving the EDCU, i. e. the car, in an unpredictable and uncontrolled situation.

Single microcomputer based EDCUs try to fulfill the safety requirements by using additional equipment such as an overspeed detection unit. This solution falls short because in case of malfunction of the control unit accidents may be caused before overspeed will be detected.

Existing control units use only one sensor for the measurement of engine speed which is the most important engine parameter. In case of failure of this sensor the engine must be stopped immediately.

Addition of functions to one-microcomputer systems often is impossible due to lack of memory space, microcomputer ports or timer/counters. In any case, however, the new functions have to be squeezed in between the existing functions which may be downgraded because any computer can only perform one task at a time.

NEW CONCEPT - BASIC FUNCTIONS

The new concept for electronic diesel engine control, designed by an affiliate to Voest-Alpine A.G., Austria, and implemented in the electronic diesel engine governor type 3 (see Figure 1 * for a block diagram and Table 1 for a summary of technical data), overcomes the shortcomings of existing EDCUs. The basic model performs the control of fuel injection quantity. The essential advantages of the new concept are as follows:

MULTI-MICROCOMPUTER ARCHITECTURE

In order to overcome single-microcomputer system limitations a multi-microcomputer architecture is employed - to our knowledge for the first time in such and similar instruments. The new concept is based upon a powerful failsafe array of 2 to 4 full-fledged third-generation 8/16 bit microcomputers (Fig. 2). The tasks are distributed among the computers and hence executed simultaneously, increasing the processing power considerably over any single microcomputer system.

This approach is different from the well-known method of operating 2 microcomputers executing identical programs in parallel and comparing their results. The method of parallel processing does not only use microcomputers in a very inefficient manner, but also does not yield any useful information in case of different results.

The performance of this multi-microcomputer concept goes far beyond. The individual microcomputers control each other irrespective of the fact that they handle different tasks; they do this, moreover, so meticulously that each deviation from the desired behaviour will be either - mostly imperceptibly - corrected or, in case this is not possible (e. g. EDCU, actuator, control rack defective) the emergency stop will be activated. The emergency stop command can be issued by any of the computers independently.

It is important to note that this is not a multi-microprocessor system, but an array of single-chip computers. The address and data buses are on-chip, they are not routed off-chip and thus fully immune to disturbances by interference. Communication between the computers and from the parameter memory is effectuated via a failsafe special bus not related to any of the microcomputer buses.

The distribution of tasks among the microcomputers also offers more flexibility relating to program modifications and to the addition of new functions.

---

* Illustrations are arranged at the end of the paper.

Program modifications mostly concern only one microcomputer program rather than the total program. This is not only easier to implement, but also it is impossible to create errors in the other microcomputers' programs. The addition of new functions is simply done by adding another microcomputer and its associated hardware and software to the system. It is also easily possible to exchange subsystems, e.g. in case of use of a different actuator subsystem.

REMOVABLE PARAMETER MEMORY MODULE

One of the principal goals was the development of a standard electronic diesel engine control unit. Therefore all engine and vehicle data and parameters are stored in a plug-in module which contains a semiconductor memory specified to withstand the severe environment in motor vehicles. In case of repair this module is removed from the defective control unit and plugged into the replacement unit. This turns out to be the only safe method which guarantees the correct transfer of engine and vehicle data to the new control unit.

ENGINE SPEED MEASUREMENT

Engine speed signals are measured from two independent sources (inductive sensor at the flywheel and generator signal) and continuously compared. In case of failure of one of them, operation of the control unit remains unaffected.

SENSORS

Current engine data such as boost pressure, accelerator pedal position, intake air, fuel and engine temperatures may be measured by different types of sensors. A major problem is the precise evaluation of analog sensor signals over the extreme temperature range of -40 to +125 degrees Centigrade and the life span of about 10 years.

The new concept eliminates measurement errors caused by the electronics over the entire temperature range and over the entire life span of the control unit. In case of the boost pressure sensor its production tolerances are compensated for such that aging only will affect the accuracy of the measurement. In case of failure of the boost pressure sensor or the accelerator pedal sensor, no boost pressure or accelerator pedal zero position, respectively, will be assumed. If a temperature sensor fails, a customer-defined default value will be used instead.

ACTUATOR AND POSITION FEEDBACK

The actuator of the electronic diesel engine governor type 3 is a stepper motor which is connected mechanically to the control rack of an in-line fuel pump or to the common control rack of unit injectors. It is driven by one microcomputer via an innovative low-cost, small and low dissipation power stage. This electronics is insensitive to short circuits and disconnection, under overload it is automatically switched off.

The position feedback system consists of an absolute, failsafe high-resolution angular position sensor mounted to the stepper motor. Absolute sensing not only is more reliable than incremental sensing, but in case of power interruptions the position can be read immediately without the need of returning the control rack to a reference position.

OVER-TEMPERATURE DETECTION

Besides reducing the number of elements, keeping the operating temperature as low as possible increases the reliability of the control unit. If the internal temperature of the control unit exceeds 110 degrees C a signal will be generated and sent to a microcomputer. At the same time a warning lamp will start blinking. Further action is up to the car manufacturer to decide, e. g. engine power may be slowly reduced which may take several minutes allowing the driver ample time to get the car out of traffic. Once the driver turns the motor off starting will be inhibited until the EDCU has cooled down sufficiently indicated by the control lamp extinguishing.

COMMUNICATION WITH EXTERNAL DEVICES

Besides the communication of the electronic diesel engine control unit with its associated programming computer it is desirable to establish communication with various external devices.

For that purpose, the control unit offers a multi-user serial interface with bus properties according to the RS-485 standard. The communication protocol is specified by the control unit. By use of an additional interface adapter, it is possible to communicate with the control unit according to other hardware standards, e. g. RS-232, or even according to other standard or customer-defined communication protocols.

NEW CONCEPT - ADDITIONAL FUNCTIONS

Future models of the governor type 3 will cover, e. g. the following functions; other type EDCUs for different actuators respective position feedback sensors are also being developed:

EXHAUST GAS RECIRCULATION (EGR)

EGR is an effective method to meet emission standards. The new concept supports an EGR control mechanism with an EGR valve and a throttle valve which are controlled independently of each other in a servo loop. The positions of the EGR as well as the throttle valves are functions of engine speed and load. These functions are represented by 3-dimensional characteristics which are stored in the parameter memory. The influence of the engine temperature on the EGR and throttle valve positions is compensated for by corrective terms.

INJECTION TIMING

Control of injection timing influences fuel consumption, engine noise and emissions positively. Similar to the exhaust gas recirculation the beginning of the fuel injection is a function of engine speed and load. The 3-dimensional characteristic representing this function is stored in the parameter memory as well. The influence of the engine temperature on the injection timing is again compensated for by a corrective term.

CRUISE CONTROL

The cruise control function requires a speed sensor and inputs from its respective control switches.

CONTROL OF AUTOMATIC TRANSMISSIONS

The EDCU supplies two signals to an electronically controlled automatic transmission. The frequency of the first signal is proportional to the engine speed. The second signal gives information about the current torque of the engine. A signal from the transmission electronics causes the EDCU to momentarily reduce engine torque while gears are being shifted.

ENGINE DATA DISPLAY

Engine and vehicle data such as engine speed, engine temperature, boost pressure, cruising speed, etc. may be supplied in an appropriate form to conventional moving-coil meters for indication on the dashboard, thus saving the car manufacturer the expense of the associated sensors, harness and electronics.

PROGRAMMING COMPUTER

During the entire engine and vehicle development phase, the engine and vehicle parameters of the electronic diesel engine control unit have to be accessible. The motor engineer must be given a handy tool to change these parameters even while the engine is running and without help of the control unit's manufacturer.

For this purpose a portable programming computer with color graphics was developed, which is connected to the control unit via the serial interface. The user is guided by a dialogue in a menu fashion, and the data is displayed on the screen in terms and formats familiar to the motor engineer. Errors caused by incorrect data input as well as error messages received from the control unit are displayed at the bottom of the screen.

The programming computer supports administration and management of complete sets of engine and vehicle parameters on a storage medium as well as modification of individual parameters. These functions can be performed either on-line, i. e. while the engine is running, or off-line, e. g. to prepare different parameter sets for test purposes. Critical engine parameters such as maximum engine speed or full fuel limitation may be entered and changed only by authorized personnel in order to prevent engine damage.

Besides these features the programming computer provides functions for collection and display of real-time engine data such as engine speed, accelerator pedal position, boost pressure, intake air temperature, control rack position, etc. for examination by the motor engineer.

Table 2 summarizes the technical data of the programming computer, Table 3 lists its detailed functions.

SUMMARY

A new electronic diesel engine control unit line overcomes limitations of similar instruments known sofar. For the first time a multi-microcomputer architecture is employed. Safe and reliable operation as well as a multitude of functions are realized.

The system includes a programming computer, a service unit for repair shops and a maintenance unit for EDCU repair centers. A multi-user serial interface is provided allowing incorporation of the unit into a network.

REFERENCES

[1] Seibt,A.: Elektronische Regelung von Dieselmotoren. 6. Wiener Motorensymposium, Vienna, Austria, 9-10 May 1985.

[2] Bittinger,W., Kainz,A., Augesky,C.: Electronic Control of Diesel Injection Systems. ISATA International Symposium on Automotive Technology and Automation, Graz, Austria, 23-27 September 1985.

[3] Bittinger,W.: Electronic Control of Diesel Fuel Injection. Fifth International Conference on Automotive Electronics. Birmingham, U.K., 29 October - 1 November 1985.

Figure 1 : Block diagram of governor type 3

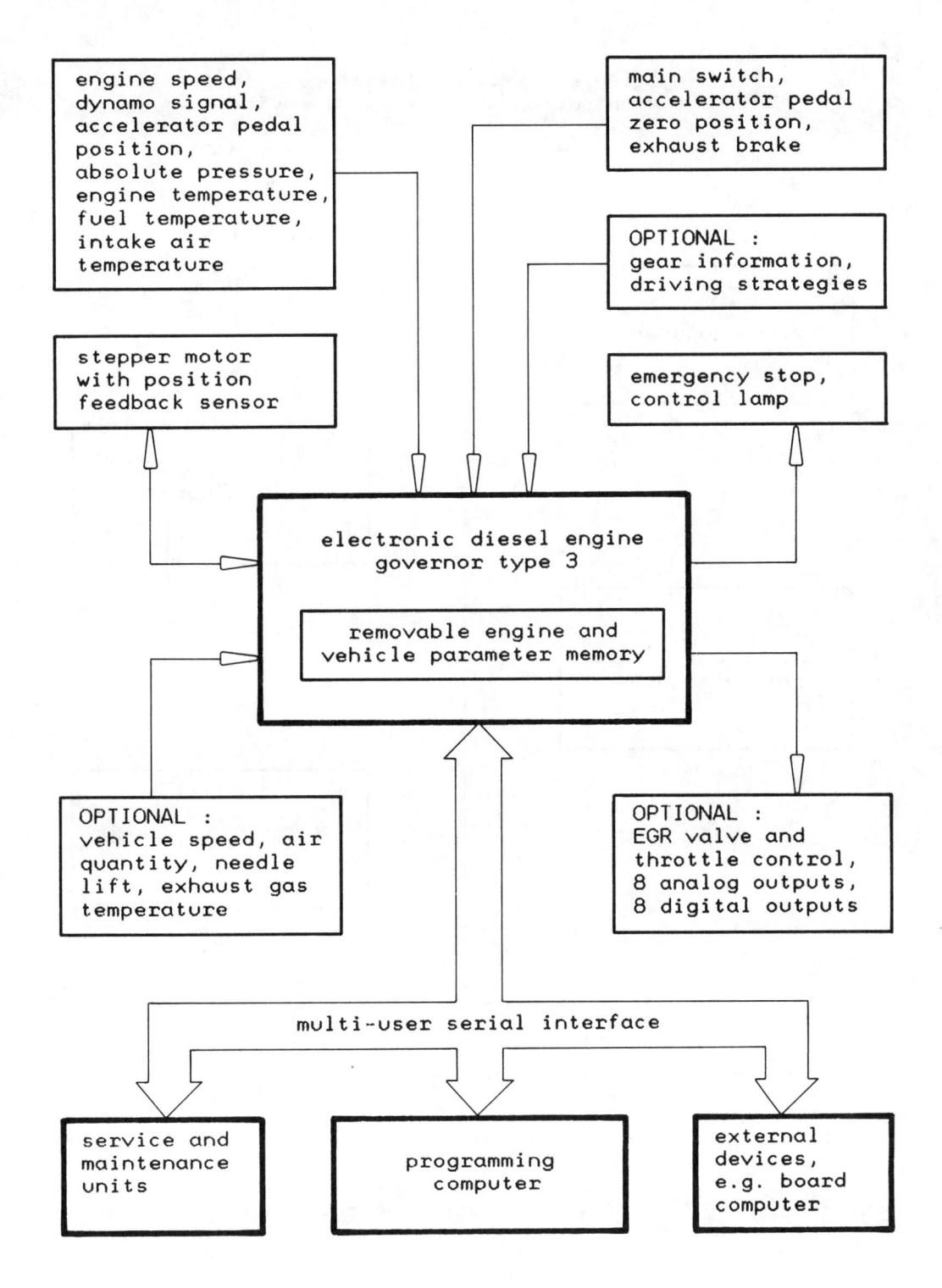

VAF 8601

Figure 2 : Multi-Microcomputer Architecture

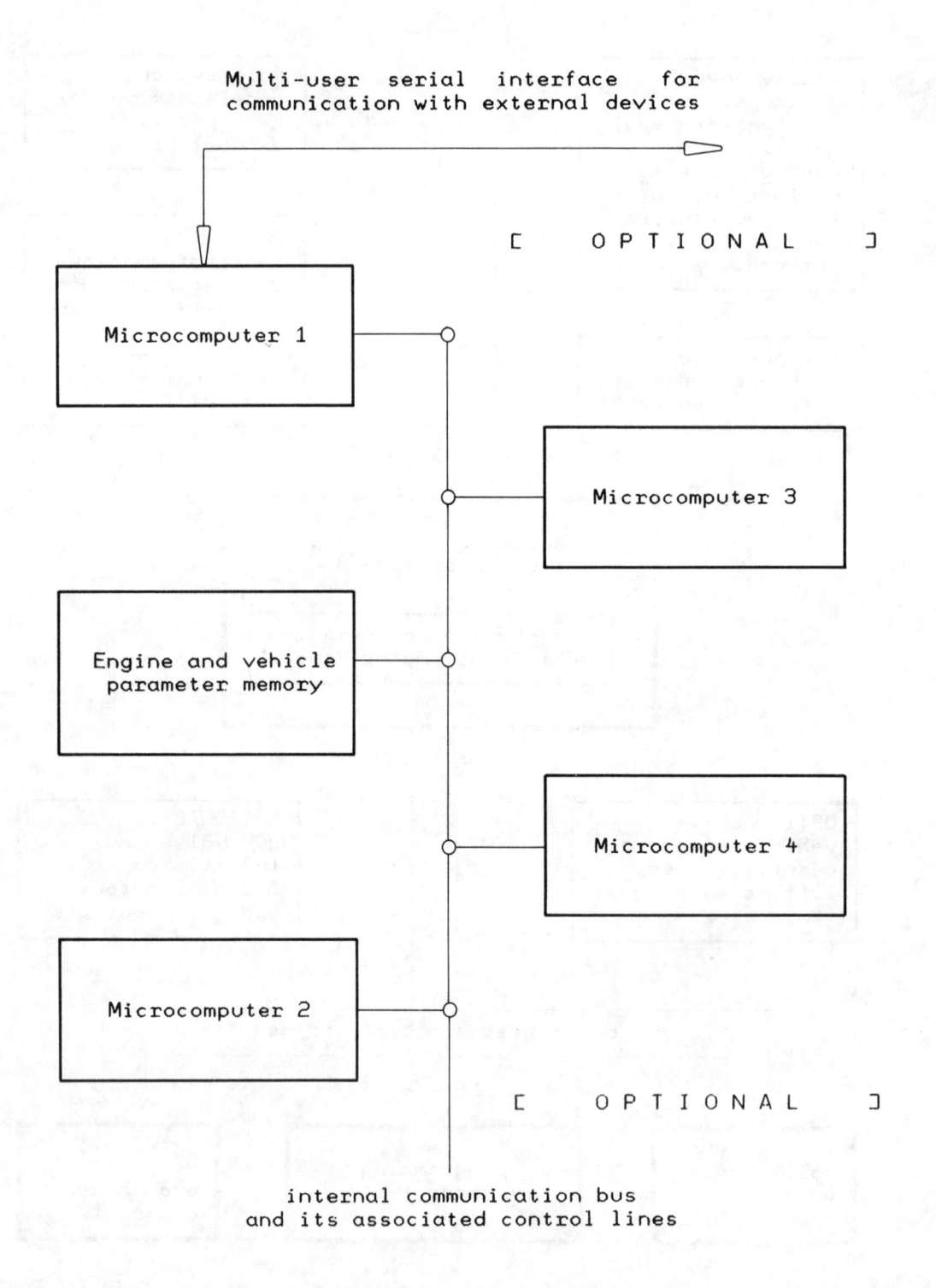

Table 1 : Summary of technical data of governor type 3

Dimensions of governor	ca. 190 x 140 x 60 mm
Connectors	1 or 2 35 pin connectors AMP, splashproof
Parameter module	Plug-in module, semiconductor memory containing all engine and vehicle data
Environmental conditions	-40 to 85 deg. C continuously, 110 degrees C for short periods (self-protective circuit automatic soft shutdown), splashproof housing
Power	12 or 24 V nominal, operating range 6 to 36 V, power dissipation approximately 6-8 W
Inputs	Main switch, zero position of accelerator pedal, exhaust brake
Optional inputs	8 inputs, e.g. cruise control, gear information, driving strategies etc.
Outputs	Control lamp, external emergency stop
Optional outputs	EGR valve control, boost pressure limitation valve, 8 analog outputs for display of temperature, boost pressure, engine and vehicle speed etc., outputs for transmission control, control lamps etc.
Sensors	3 temperature sensors (operating range -40 to 150 degrees C), 1 absolute pressure sensor (barometric or boost pressure, operating range 600 to 2000 mbar), 1 accelerator pedal position sensor, 1 engine speed sensor (inductive type, up to 20 kHz), dynamo signal
Optional sensors	Air quantity measurement, exhaust gas temperature, vehicle speed, needle lift sensor
Stepper motor	200 steps per revolution, -40 to 150 degrees C operating temperature
Position feedback system	Absolute position indicator, 200 steps per revolution, -40 to 150 degrees C operating temperature
External serial interface	Type RS-485 interface for programming computer and service and maintenance units; allows on-line connection of an arbitrary number of external communication partners

VAF 8603

Table 2 : Summary of technical data of programming computer

Dimensions	ca. 470 x 195 x 540 mm
Screen	165 x 102 mm, high-resolution color graphics monitor, connector for external color graphics monitor
Keyboard	Standard keyboard with pre-programmed menu keys
Serial interfaces	1 serial interface according to RS-485 and 1 serial interface according to RS-232 standard
Main memory	256 kbyte RAM, expandable with RAMs or EPROMs up to 1024 kbyte
Background store	Optional 2 x 5.25 inch floppy disk drives with each 700 kbyte
Environmental conditions	0 to 50 degrees C
Power	220 (110) V nominal, power dissipation approximately 80 W

Table 3 : Functions of programming computer

OBJECT	OPERATION

Parameter set administration	1.) Display summary of present parameter sets on screen or print out
	2.) Delete one existing parameter set
	3.) Copy one parameter set
	4.) Compare two parameter sets listing the results on screen or paper
	5.) Modify a parameter set
	   a) on line :
	      Receive parameter set from governor or transmit parameter set to the governor
	   b) off line :
	      Restore parameter set from memory or store parameter set into memory

VAF 8604

Table 3 (Cont.) : Functions of programming computer

OBJECT	OPERATION
Modification of an individual parameter	1.) Edit a parameter value 2.) Insert a new parameter value 3.) Delete an existing parameter value 4.) Get information about the parameter on screen or paper 5.) Display parameter in graphic representation on screen 6.) Accept parameter modification (transmit parameter to governor or store it into memory)

Real-time data operations	1.) Specify engine data to be evaluated
	2.) Display specified engine data of the running engine on screen
	   a) as text and numbers
	   b) in graphic representation
	3.) Collect specified engine data of the running engine in memory
	4.) Specify evaluation conditions
	5.) Evaluate collected data according to the specified conditions and display the results on screen or print out

Error message display	1.) Display programming computer internal error messages due to incorrect data input 2.) Display error messages received from control unit

VAF 8605

860142

Exhaust Emissions Influenced by Electronic Diesel Control

E. Gaschler
Res. Div.
Volkswagen AG
W. Germany

SUMMARY

The start of injection and fuel quantity can be optimally suited to a given Diesel engine and its operating conditions by using electronic injection feedback control.

In this way, it is possible to control exhaust emissions, i.e. to reduce the emission of individual pollutants.

An optimization based on an engine map projection for a specific driving cycle helps to accelerate the design process in the engine testing bed under observance of predefined limits; this fact is confirmed by checks on the vehicle as a whole.

1. INTRODUCTION

Pollutant minimization is presently the focus of further development in automobile technology.

The drastic reduction of pollutants is the declared goal of a number of European governments, especially that of the Federal Republic of Germany.

Both petrol and Diesel vehicles are equally affected by these plans, even though the components of their pollutant emissions are of differing levels of concentration.
The chances for the present design of Diesel engine with the usual swirl and prechamber combustion process in passenger vehicles of complying with the US limits are not bad, naturally dependent upon the respective test weight class. However, additional measures are just as necessary in this case as e.g. the catalyst is for the petrol vehicle.

Alongside those pollutants limited by legislation, namely carbon monoxide (CO), hydrocarbons (HC) and nitrogen oxides (NO_x), the Diesel vehicle produces particulate emissions, which as from 1987 will be limited in the USA to 0.2 g/mile and will cause great problems especially for the larger classes of vehicle.

With the Diesel engine, the black smoke produced under full-load must likewise also be taken into account, the limits of which have frequently been pushed too high in favour of greater torque and thus lead to the black clouds which can be especially observed with vehicles climbing inclines.

2. MAIN FOCUS OF OPTIMIZATION

Since the old conflict of aims - the conflicting behaviour of the engine with regard to consumption minimization and improvement of exhaust emission quality - still makes all research and development work on the Diesel engine more difficult, variables of such different natures must be balanced against each other.

Due to the importance attached to the raw material situation, the focus in past years lay clearly on the optimization of fuel consumption. Although a steady improvement of the pollutant spectrum as a whole was achieved, tuning engines for optimum consumption lead, however, to the levels of some pollutants being so high that limits (e.g. US 75) could not be observed.

Today and in the future, therefore, the emphasis will be on tuning the engine to obtain low pollutant emission levels, whereby, however, (conflict of aims) a rise in consumption must once again be accepted.

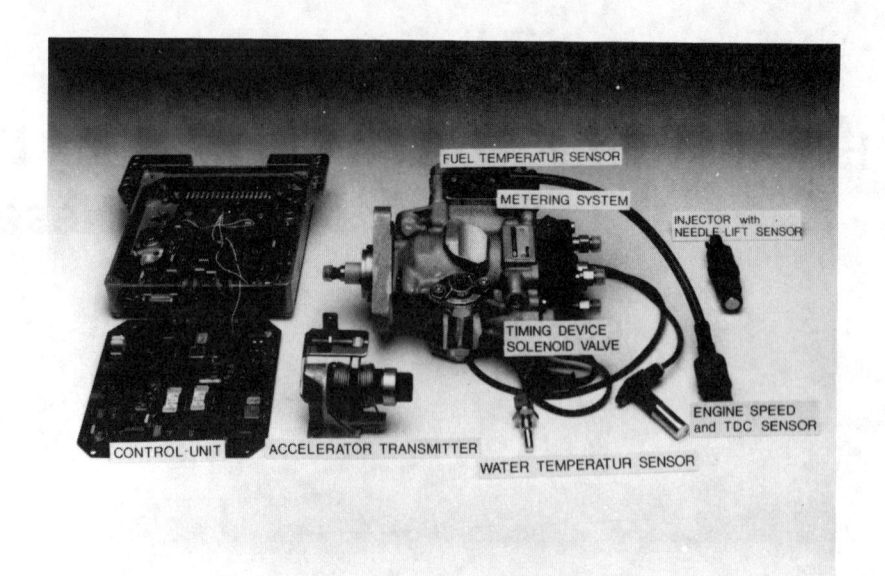

Fig. 1: Electronic Injection Feedback Control

3. ELECTRONIC INJECTION FEEDBACK CONTROL

Alongside the improvements to the exhaust emissions of the Diesel engine through internal, i.e., combustion techniques, the injection equipment - the central components of which are the pump, line and injector - offers a potential for innovation with regard to improved pollutant emission levels.

Emissions and consumption depend amongst other things to a great extent on the injection timing. However, this can only be approximately adjusted to suit the engine conditions throughout the whole operational range with the mechanical injection pump which is fitted nowadays. Similar circumstances exist for the metering of fuel.

As a part of the injection equipment, the electronic injection feedback control (EDC = Electronic Diesel Control) offers a good possibility of successfully countering the above-mentioned problems.

Figure 1 shows such a system with all the accompanying components.

The main features are the electronically regulated injection controller (injection timing) and the fuel quantity controller. With the aid of the control unit, both parts of the pump are regulated according to a definite programme, in which engine consumption maps for a particular application are stored. In Fig. 2 the emissions to be controlled are represented in tabular form.

Control of Fuel Quantity

- Full Load - Smoke
- Particulates

Control of Start of Injection (S.o.I.)

- Gaseous Emissions   - NO_x
                                     - HC
                                     - CO
- Particulates

Fig. 2: Exhaust Emissions influenced by EDC

Fig. 3 shows the control of full-load smoke emissions through quantity feedback control.

Contrasted with each other are the curves of the smoke emission of a Diesel without electronic feedback control and the same engine with electronic feedback control.

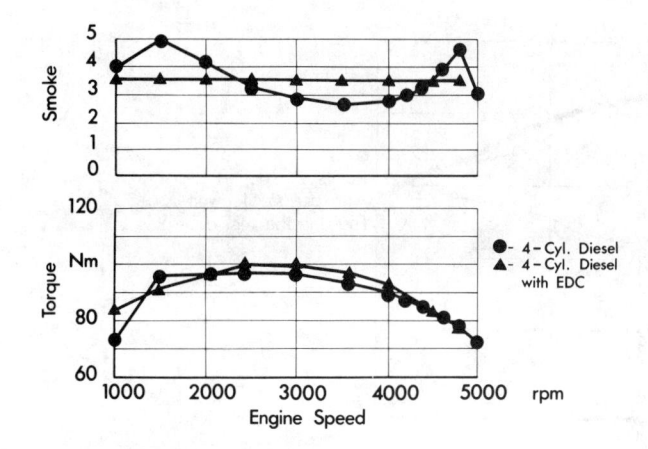

Fig. 3: Smoke Emission and Torque (full-load)

The engine with electronic injection feedback control can be driven throughout the whole engine speed range with constant smoke emission (in this case 3.5 Bosch). This means a reduction of the high smoke peaks and an extra injection of fuel at those points where reserves in the smoke emission exist, such as here in the middle engine speed range.

A secondary effect of this constant full-load smoke limit is shown in the lower part of the diagram in Fig. 3.

A gain in torque can be achieved in a wide engine speed range. At 1500 rpm a minor loss must be accepted, a loss which arises through the reduction of fuel quantity because of the high smoke emission at this point.

The more important part of the electronic feedback control at this time is the start-of-injection-control, as shown in Fig. 2. The significance for the limited pollutants CO, HC and NO_x will be illustrated on the basis of the US City Test mentioned at the beginning.

Fig. 4 shows the US 75 Test and the limits thereof.

This test begins with a cold drive-off, i.e. with a cold engine. The test has three phases. The third phase corresponds to the first, but differs in that, despite a break of 10 minutes, the engine is at its normal operating temperature.

The average speed amounts to 38.6 km/h without idling phases, the latter amounting to 18% of the total test. The essence of this simulated urban drive is acceleration, deceleration and idling phases.

Fig. 5 shows the corresponding loads for this test in an engine map for a 1.6 litre naturally aspirated Diesel Golf.

Taking the pollutant NO_2 as an example, ranges are given here showing to how many per cent of total NO_x is represented by this pollutant.

It can be determined, e.g., that the 93% range takes in less than 1/3 of the total operational engine map.

Through the electronic start-of-injection-control it is possible to meet quite specific requirements, e.g., to so set and store the start-of-injection points that the requirement

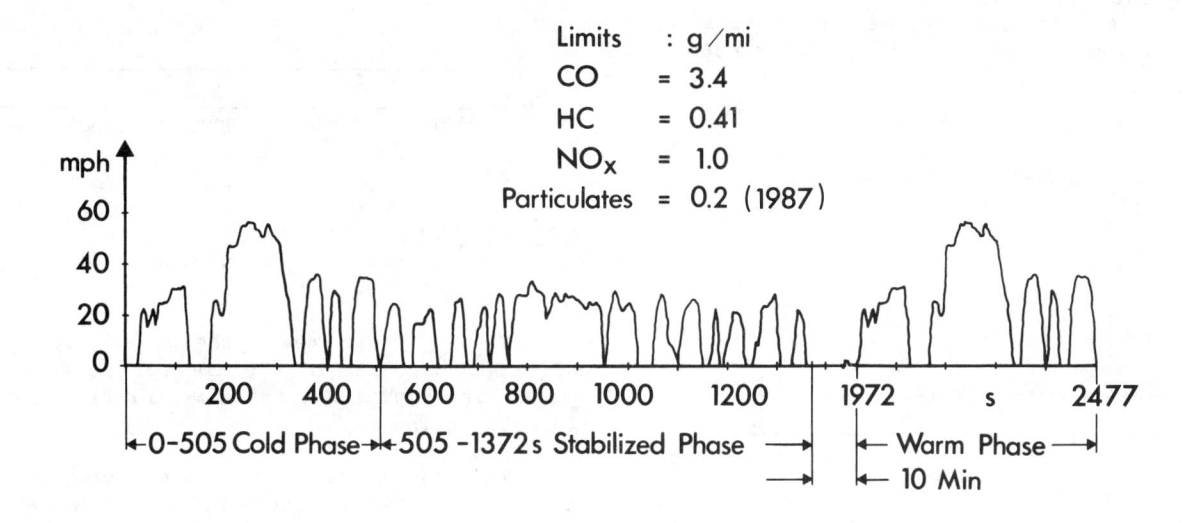

Limits : g/mi
CO = 3.4
HC = 0.41
NO_x = 1.0
Particulates = 0.2 (1987)

Fig. 4: US City (Exhaust test) Cycle

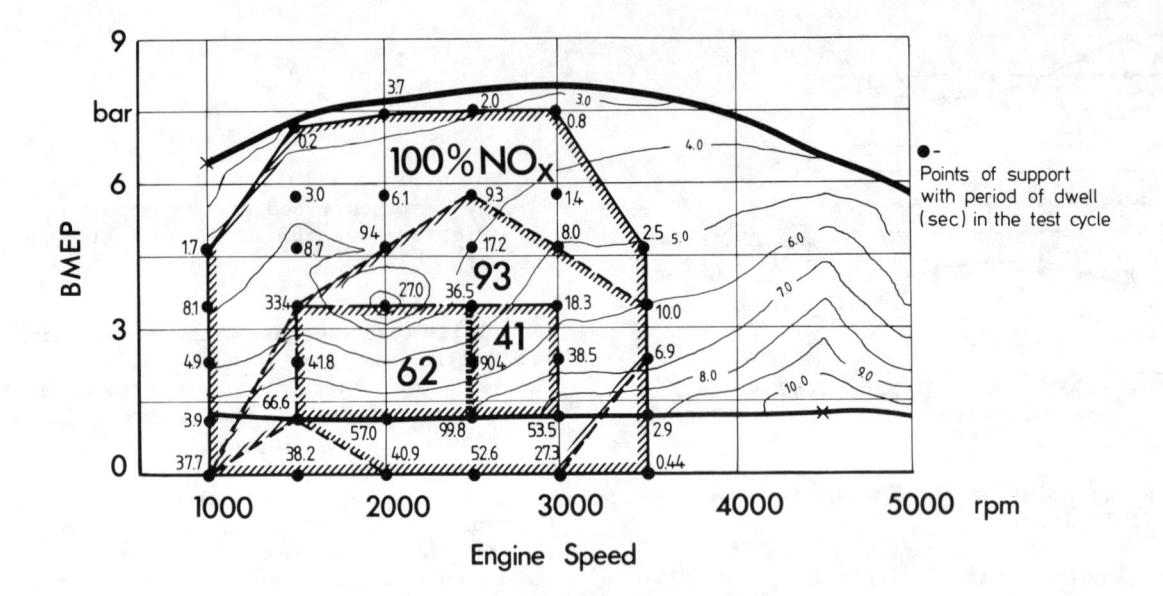

Fig. 5: NO$_x$ engine map with loads for US 75 Test. NO$_2$ (g/kWh)

for different ranges of an engine map is fulfilled.

In concrete terms, the start-of-injection could be set, e.g., in the 93% range, in such a way that the formation of NO$_x$ is reduced. The rest of the map ranges could be programmed for minimized consumption and the start-of-injection points be set at full-load in such a way, that full torque and desired rated output are achieved.

Two extreme start-of-injection engine maps are presented as examples in Figures 6 and 7. Fig. 6 shows a map, in which the start-of-injection points were consistently optimized to obtain minimal specific consumption.
In Fig. 7 a map is shown, in which optimization was carried out for NO$_{x\ min}^{*}$ below 6 bar mean pressure, i.e. relatively retarded start-of-injection points, which are rapidly advanced in the range from 6 bar mean pressure to the full-load curve.

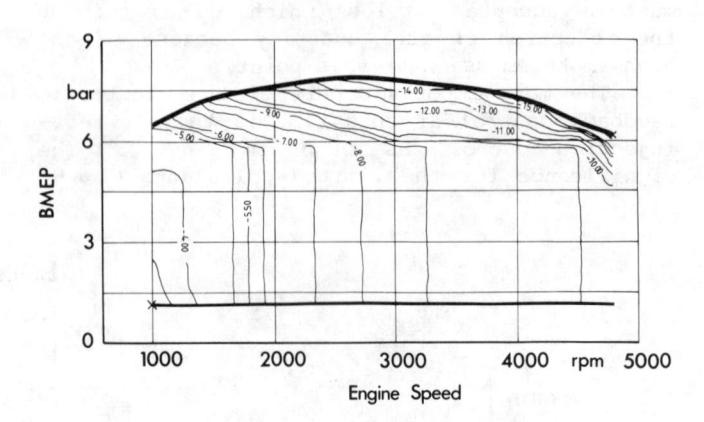

Fig. 7: Start of injection map. SoI selection according to NO$_{x\ min}^{*}$. SoI (°BTDC)

NO$_{x\ min}^{*}$ here means the lowest achievable NO$_2$ concentration in the control range of the pump. The engine has the same torque and output in both engine maps.

The effects of these extreme programmes on the pollutants and consumption are shown in Fig. 8. The values are projected with the aid of the loads represented in Fig. 5.

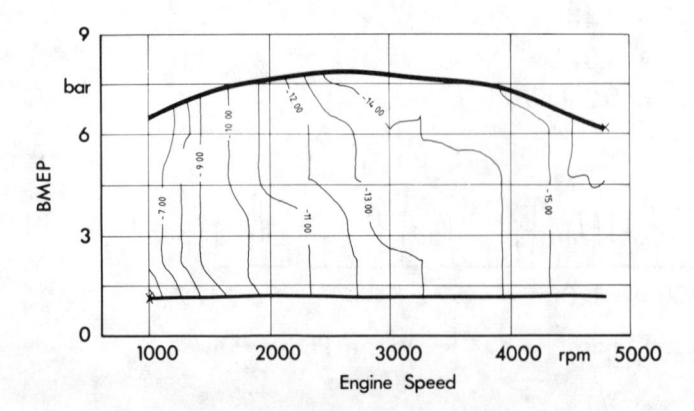

Fig. 6: Start of injection map. SoI selection according to be$_{min}$. SoI (°BTDC)

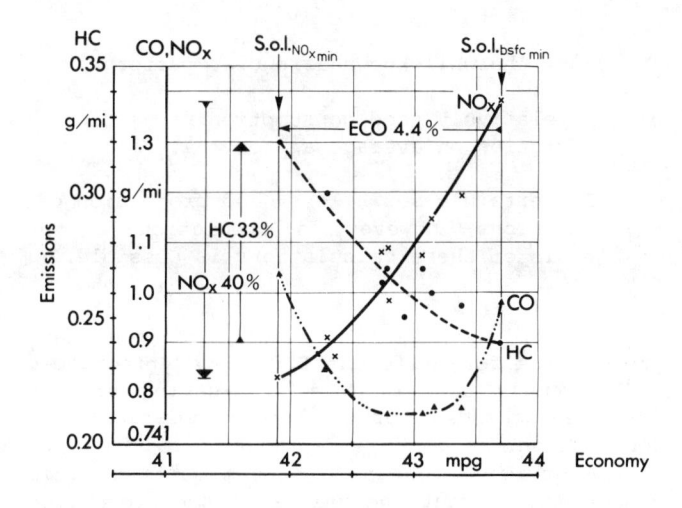

Fig. 8 Influence of the start of injection on emissions and consumption in the US 75 Test (Projection)

A comparison of the consumption- optimized engine (SoI-be$_{min}$) with the NO$_x$- optimized engine (SoI-NO$_{x_{min}}$*) shows that a NO$_x$ reduction of 40% is possible, while HC rises by 33% and the fuel consumption rises by only 4.4%. The CO level lies way below the limit.

Between the two limit programmes, all other combinations are possible.

4. OPTIMIZATION ACCORDING TO LAGRANGE

It proves to be relatively simple to select start-of-injection points according to minimal specific consumption. However, this method leaves to chance the level of emissions which arise and whether, the US 75 Test will be fulfilled or not.

If we abandon the method of selection according to be$_{min}$, due to the fact that the critical NO$_x$ limit is not fulfilled, we are then obliged to use a trial and error method, for as long as it takes till the desired NO$_x$ value is achieved. This takes time, because new programmes to be stored must be made and tests must be run again and again.

The "minimum-search" method according to Lagrange is available to help to solve this classic problem of optimization theory in a simple manner.

A summary variable must be minimized while adhering to parameters (Fig. 9), i.e. a minimum value for the components CO, HC, NO$_x$ and consumption (V) must be sought; the limits of the pollutant levels represent the parameters.

Lagrange - Optimization

$$V + L \cdot NO_x + L \cdot HC + L \cdot CO = \text{Minimum} \quad (1)$$

$V \Rightarrow \text{Min}$
$NO_x \leq \overline{NO_x}$
$HC \leq \overline{HC}$
$CO \leq \overline{CO}$

L = Lagrange - Multiplier
V = Test Cycle-Fuel Consumption
$\overline{NO_x}$, $\overline{HC}$, $\overline{CO}$ = Limits

Fig. 9: Optimization according to Lagrange

The advantage of the general minimum-equation (1) as presented in Fig. 9 is that an optimization rule can be derived from it for each separate operating point (base point) /2/, although the limits for the whole driving cycle are defined. The selection of components becomes less if completely uncritical ones can be omitted, e.g. CO.

The practical meaning of this process is that we increase the Lagrange multiplier L in a projection based on a driving cycle to be determined with a relatively simple calculation programme from L = 0 (optimum consumption) until the desired emission limits are achieved.

5. RESULTS ON THE DYNAMOMETER

The measurements on the dynamometer were interesting in two respects.

o Establishing the actual consumption and emission values for the 4 cylinder Diesel with electronic feedback control with different stored start-of-injection points in comparison to the engine without EDC.

o Comparison of the values from the projection process which represented the basis for all start-of-injection optimizations, with those of the dynamometer measurements.

Fig. 10 shows the measured emission and consumption values.

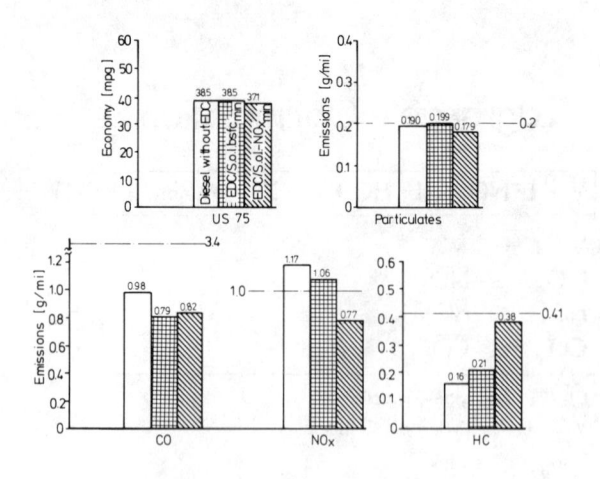

Fig. 10: USA: Economy and Emissions: - Diesel without EDC
- Diesel with EDC
(Test vehicle: Golf MJ 84)

The measurements on the dynamometer confirm the projections. They also confirm and show the possibilities to control the composition of emissions via different optimizations using start-of-injection points, which are only possible in this form with electronic Diesel feedback control.

The critical NO_x emissions could be reduced from 1.17 g/mi to 0.77 g/mi with an increase of hydrocarbon emissions to 0.38 g/mi.

The particulate emissions, which until now have not been mentioned, because they are not arithmetically ascertainable from the smoke values, also show a downward tendency in the NO_{xmin}*-optimization.

The deterioration of consumption or economy loss from 38.5 mpg to 37.1 mpg for the EDC engine with NO_{xmin}*-optimization is at 4.4% exactly within the predicted range.

It should be noted that these tests are concerned with individual engines and that dealing with full production scatter requires additional efforts.

A comparison of the projected values with those of the dynamometer is presented in Fig. 11 for two different start-of-injection optimizations.

They show the well-known irregular pattern.

o A relatively good consumption projection. Deviation on average approx. 7%.

o In certain cases a less good projection of emissions; however, a prediction on the basis of these calculations is possible.

The "minimum search" method based on these projections for different EDC stored programmes is extremely useful as a time-saving process for pre-assessment of levels of emissions and consumption.

A striking feature of the results of this comparison is that the values of the projection are considerably higher. It can therefore be assumed that we are erring on the side of safety when optimizing emission-related parameters.

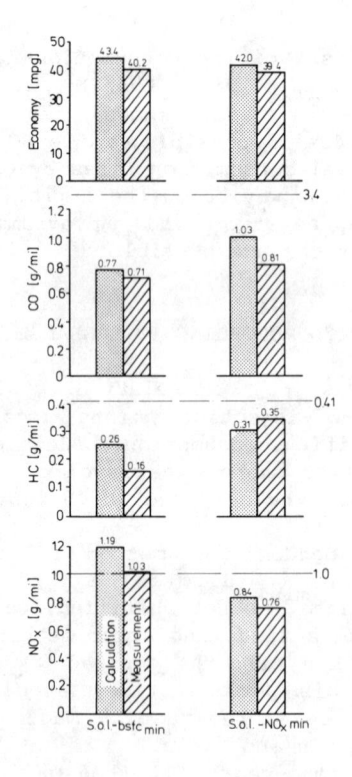

Fig. 11: US 75 Test (warm start) Comparison: Projection - Dynamometer measurement

Optimized for Minimum Fuel Consumption in the Test-Cycle

Optimized for Optimum Fuel Consumption in the Test-Cycle with Keeping all Emissions-Limits

Fig. 12: Chip change

6. CONCLUDING REMARKS

The electronic Diesel control (EDC) proves in several respects to be a suitable instrument for adapting the Diesel engine in its present stage of development to meet the requirements of the future.

This is mainly due to the freedom which this system offers, with the help of electronics, to use a change of chip (Fig. 12) to bring engines (same number of cylinders) for the most differing applications, with largely the same equipment, into an optimum condition in each case.

It is thus possible to cover not only the most widely differing driving cycles with varying pollutant limits, but also naturally aspirated and supercharged engine variants and those with luxury features.

This technology will, however, only then come into practical operation, when vehicles with conventional mechanical systems have reached the end of their capabilities and can, for example, no longer fulfil future limits or other demands.

Acknowledgement

The author would like to thank Robert Bosch GmbH (Stuttgart / W. Germany) for the good cooperation in this project, especially Mr. R. Dorenkamp and Mr. H.-G. Hummel for their efforts in adjusting the injection equipment and in preparing the software of the electronical system.

References.

/1/ H. Eisele, "Electronic Control of Diesel Passenger Cars"
SAE 800 167

/2/ B. Böning, W. Richter, W. Tuleweit and B. Zeilinger
"Systematische Optimierung von Kennfeldern für Gemischbildung und Zündung bei Ottomotoren".
MTZ 44 (1983).

860143

Digital Self-Calibrating Hall-Effect Sensor for Electronically Controlled Engines

Michael S. Ziemacki and George D. Wolff
Wolff Controls Corp.

ABSTRACT

A new Hall-effect IC digital injection sensor has been developed by Wolff Controls Corporation. The sensor is self-calibrating and will be used for instantaneous fuel consumption indication and for timing and fuel input feedback on closed loop electronically controlled diesel engines. It is the only needle lift sensor giving both, a beginning of injection and a fuel quantity signal.

This paper focuses on the fuel quantity signal, its derivation and accuracy, and discusses applications using different types of injectors and injection systems.

IN ORDER TO MEET the ever increasing demands of fuel economy and emission standards, diesel engines will soon require electronic controls. Many of the techniques can be borrowed directly from work on gasoline engines but the fuel system will require many new innovations.

Wolff Controls began development of Hall-effect needle lift sensors for diesel injectors in early 1979. The Analog Needle Lift Sensor is now widely used worldwide for diesel engine research. Early prototypes of electronically controlled fuel injection systems have utilized the Analog Sensor for providing a beginning of injection signal (1)*. The output of the Analog Sensor is not particularly suited for the requirements of a microprocessor control system. Therefore, it was necessary to process the signal to obtain an output suitable for the microprocessor. Along with the requirement of providing a beginning of injection signal (2), it soon became obvious, as work progressed on fuel control systems, that some indication of engine load was desirable (3, 4). Other studies suggest "that the beginning and end of injection play a major role in solid particulate formation while the overall rate of injection is of secondary importance" (5). In fact, even before the work on fuel controls, a means for determining instantaneous fuel consumption for use with trip computers had been sought for dieselized luxury automobiles.

All these considerations initiated, in early 1983, the development of a new Timing Sensor at Wolff Controls Corporation. It provides both a beginning of injection signal as well as a duration of injection signal compatible with a microprocessor based control system. Fig. 1 shows the output of the Timing Sensor as compared to a needle lift trace. The current pulse output of the sensor corresponds to the beginning and end of the injection event.

In this paper, the discussion is limited to the duration of the injection event as indicated by the pulse width of the sensor signal, and how it relates to the instantaneous fuel delivery. Information on the beginning-of-injection event and other sensor aspects can be found in the references (6, 7, 8).

---
*Numbers in parentheses designate references at the end of paper.

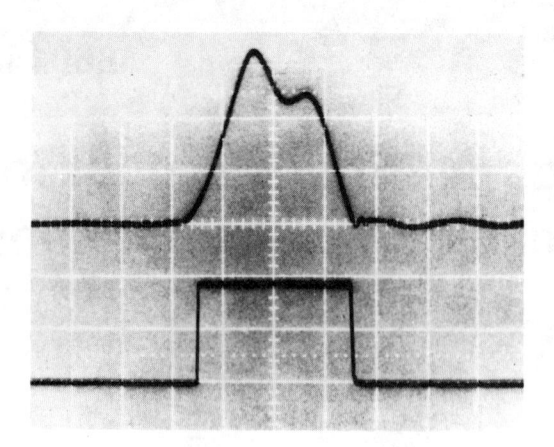

Fig. 1 - Needle lift and sensor signal

Analysis of a diesel fuel injection system is very difficult due to the complex nature of the hydraulic pulse phenomena. Fuel quantity analysis is even more difficult due to the varying throttling effect of the nozzle. In fact, most computer analyses only model the value as a constant (9). Fig. 2 shows a pressure trace along with the corresponding needle lift. It is obvious from the traces that an exact fuel quantity relationship would be very difficult to predict from the pressure trace alone. The question becomes, "can fuel quantity be derived from the needle lift signal, in particular, from the duration of needle lift?"

In order to be useful, the sensor's output must be applicable to both direct and indirect injection systems. Since most automotive applications to date have been indirect and utilize a pintle type injector, this system was first analysed.

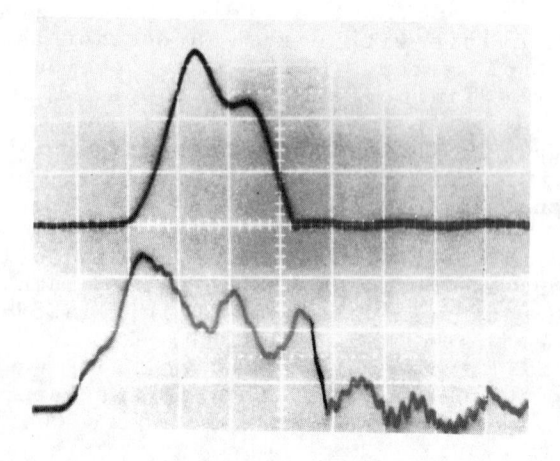

Fig. 2 - Needle lift and line pressure

PINTLE INJECTOR

Since the pintle injector needle usually does not open against a stop, the needle lift trace gives an indication of injection pressure by the amount of lift. This is not the case with hole type, direct injection, nozzles which normally reach a stop. Therefore, the trace height for the hole type nozzles gives no pressure information. To take advantage of the pressure information inferred in the pintle needle lift trace, data was taken using the area under the needle lift curve. In order to directly compare the integral data with the output from a Timing Sensor, an injector was instrumented with both an Analog Sensor and a Timing Sensor. Fig. 3 shows the configuration.

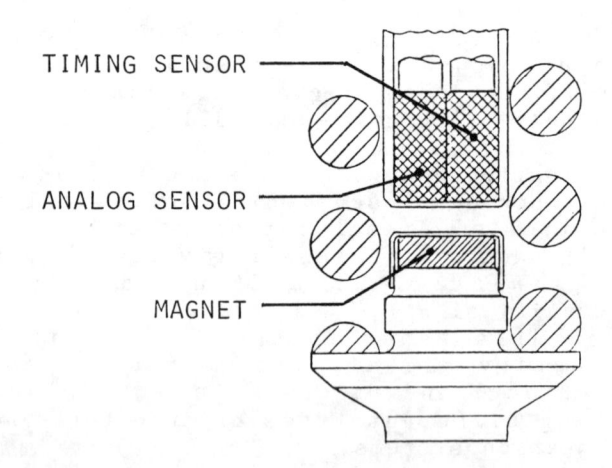

Fig. 3 - Dual sensor installation

A Norland Digital Oscilloscope was used to automatically perform the integration of the analog needle lift trace. This integral data it plotted in Fig. 4 as a function of RPM and fuel quantity. All RPM referred to in this paper are engine RPM. Data was consistent with the exception of near maximum output data. Small increases in fuel quantity were sometimes measured as a small decrease in the integral. A change in the shape of the trace was determined to be the cause of the inconsistencies.

Fig. 5 shows the Timing Sensor output data. The duration of injection in milliseconds is plotted against the fuel quantity. Time has been taken as a reference in place of crank angle since microprocessor inputs would normally be in time units, i.e. clock cycles. The areas of inconsistency noted with the integral method were not present with the duration signal.

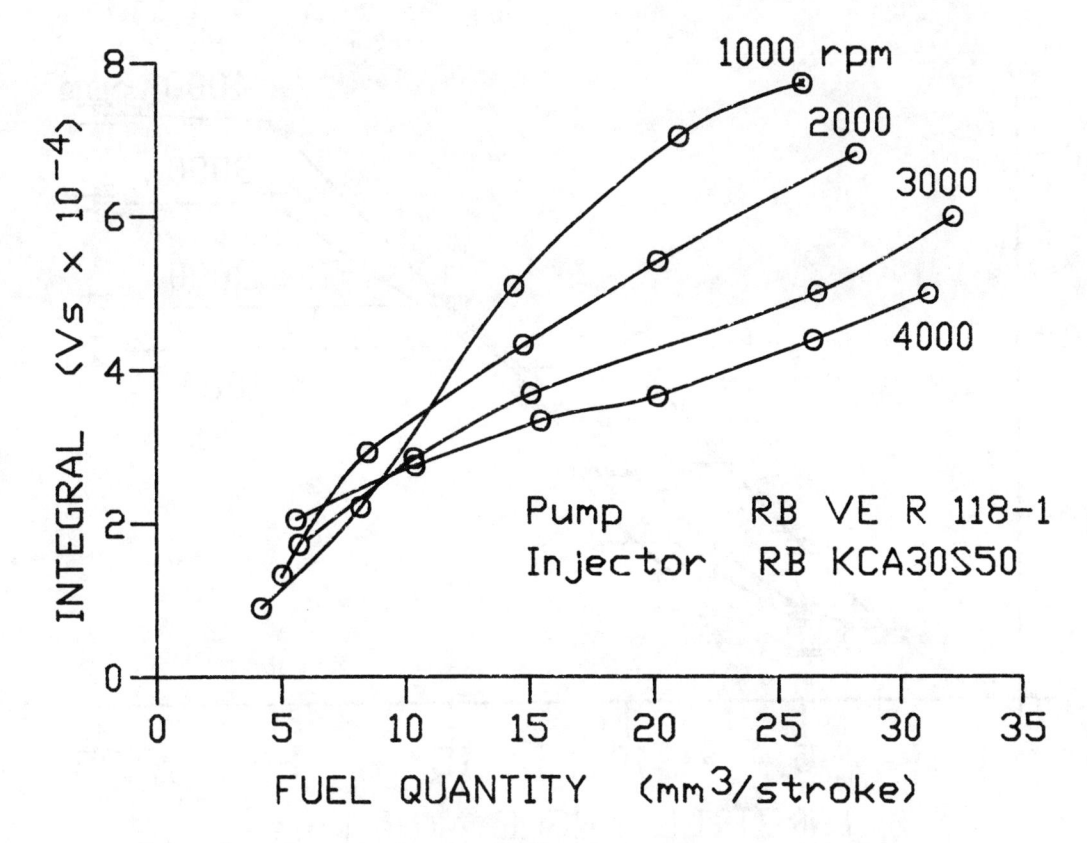

Fig. 4 - Integral of needle lift vs fuel quantity

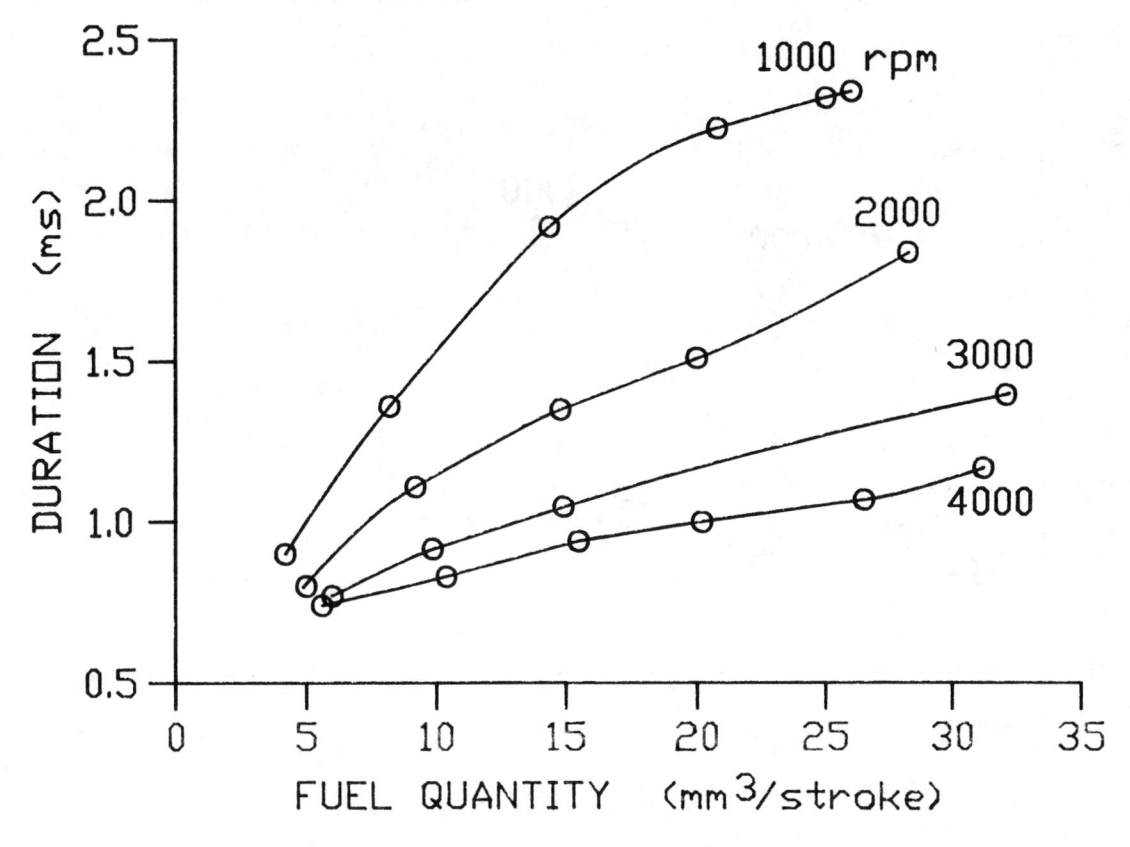

Fig. 5 - Timing sensor signal duration vs fuel quantity

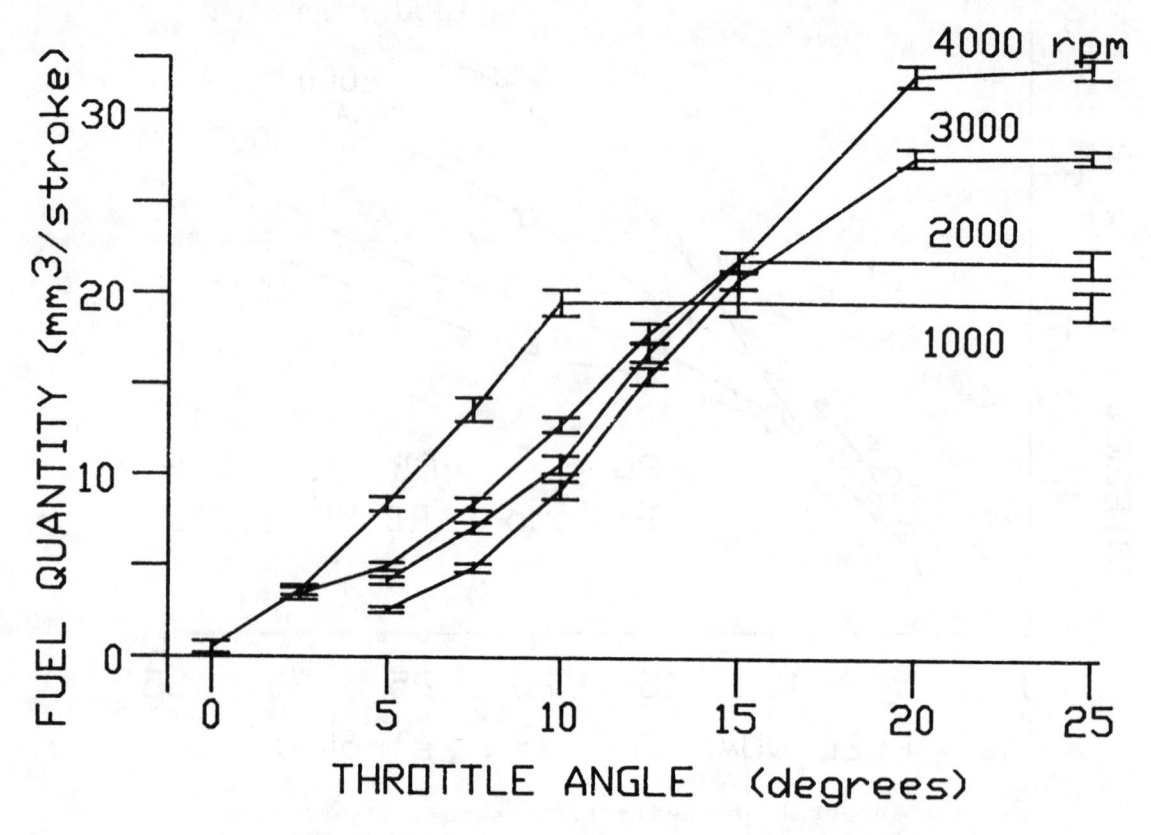

Fig. 6 - Fuel delivery vs throttle angle

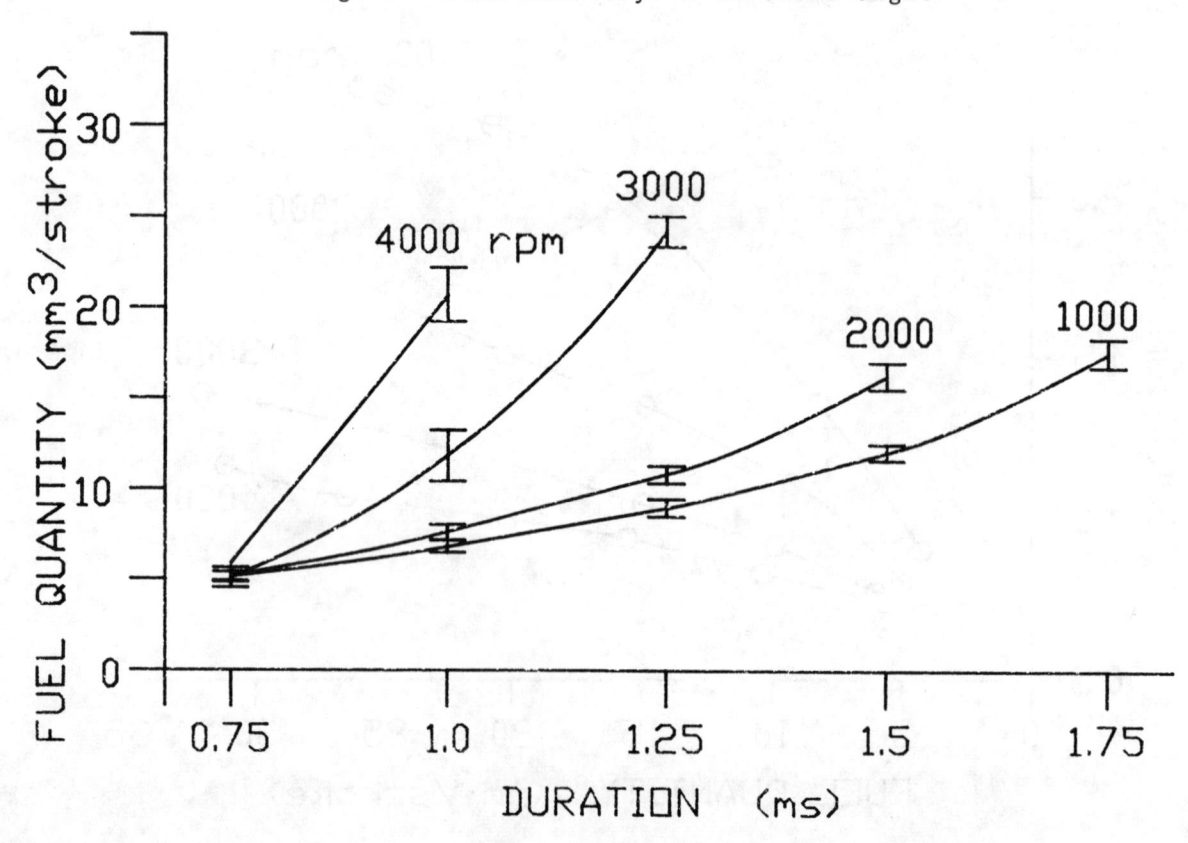

Fig. 7 - Fuel delivery vs sensor signal duration

A comparison of the curves in Figs. 4 and 5 shows very similar results. The difference is that the integral signal has less dependence on RPM. The maximum pressure is almost linearly dependent on RPM. Therefore, the integral signal, which is related to lift and in turn maximum pressure, already contains a component of RPM. The duration signal contains no such component and therefore is more dependent on RPM. The slope of the curves is inversely proportional to RPM. Fortunately, the decrease in slope at high RPM's is accompanied by an increase in the slope of the needle lift trace. This helps to maintain a consistent degree of accuracy in determining fuel delivery from duration throughout the speed range.

In order to evaluate the consistency of the duration signal, several sets of injectors were instrumented with Timing Sensors. The spring seat, which had to be modified to accept the magnet and cap, was the only part changed that affected the sensor to magnet gap. All other parts maintained standard mechanical tolerances. Tests were carried out on a pump test stand using two distributor type injection pumps of the same execution. To ensure the consistency of the data taken, fuel quantity was measured for 1000 strokes. During the same time period, each injectors duration signal was averaged for 100 events. This permitted duration data collection during the fuel quantity measurement while removing injection to injection event differences by averaging.

Fig. 6 shows the variation in fuel quantity for one pump and twelve injectors and the standard deviation for each point. The differences between injectors correlate with the opening pressure of the individual injectors. Data presented later shows the effects of opening pressure. Fig. 7 shows the variation in the fuel quantity versus the sensor signal duration for the previous series of tests. Sensor performance is hard to evaluate from this data, since it contains both injector and sensor variations. In order to gain a better understanding of the variables that affect sensor performance other tests were run.

In order to determine the effect of Timing Sensor threshold on the duration signal a series of tests were run with sensors whose threshold greatly exceeded the normal production threshold tolerance. Fig. 8 shows the change in the duration signal for six injectors. Operating conditions were: 2500 RPM, half load (14.8 mm^3/stroke).

Additional data was taken for one injector at three opening pressures to determine what effect the opening pressure had on the duration signal. Data was taken at 120, 150 and 180 bar. The nominal opening pressure is 150 ±5 bar. Nominal data, for the low and high delivery operating points, is underlined in the table, Fig. 9. The change in injection quantity is significant, nearly matching the change in opening pressure, ±20%. The column, "Fuel Estimate", shows the predicted fuel delivery based on duration signal data taken at nominal opening pressure for the respective engine speeds. While the estimate does not predict the correct fuel delivery, it is in most cases twice as accurate as the nominal reading based on throttle position.

HOLE-TYPE INJECTOR

Early consensus was that the hole type nozzles and increased pressures used in direct injection engines would not lend themselves to the duration of injection principle of determining fuel quantity, particularly, since the needle is normally going against a stop and no pressure information is contained in the trace. Typical inline injection pump data sheets, Fig. 10, do show however a linear relationship between fuel quantity and rack position or cam angle. Since the cam angle is related directly to the injection duration, there was hope for using the duration signal for determining fuel delivery.

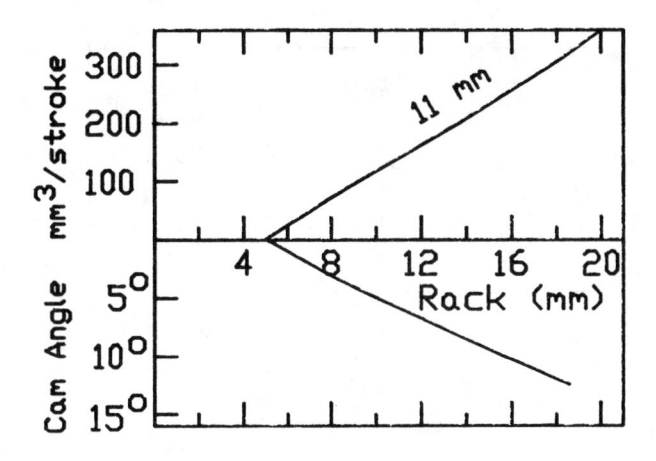

Fig. 10 - Inline pump characteristics

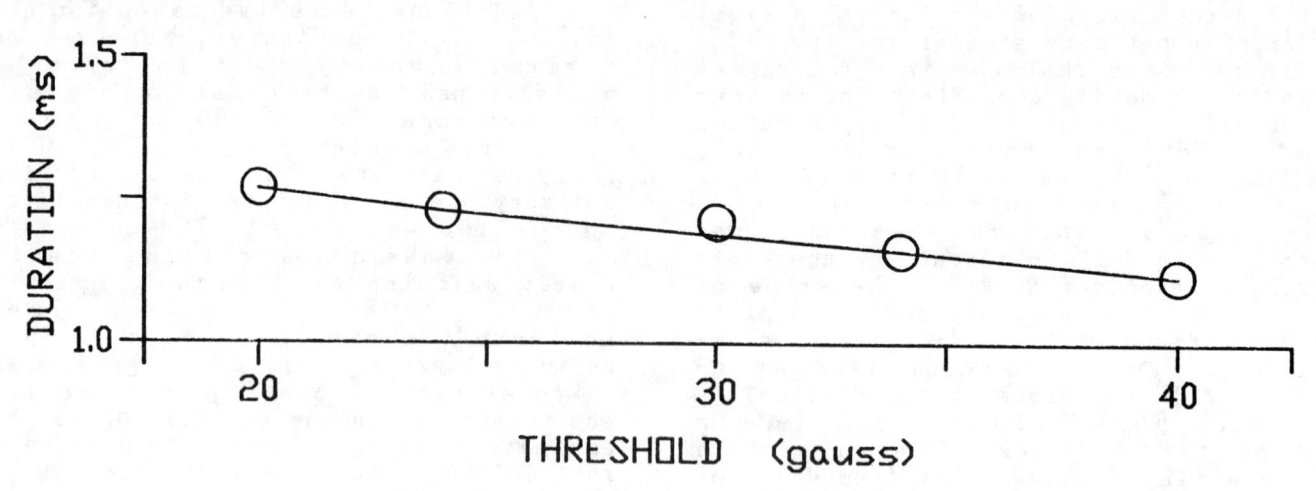

Fig. 8 - Sensor threshold influence

Engine RPM	Opening Pressure (bar)	Duration (ms)	Fuel Delivery (mm^3/stroke)	Fuel Estimate (mm^3/stroke)	Throttle Angle (degrees)
1000	120 150 180	1.42 1.11 0.73	9.0 7.3 5.2	10.5 4.8	5
	120 150 180	2.09 1.85 1.65	23.0 19.5 15.4	25.0 14.3	10
2000	120 150 180	1.50 1.37 1.16	13.6 11.4 9.8	15.1 9.3	10
	120 150 180	1.77 1.65 1.54	25.3 22.0 18.3	28.0 16.0	15
3000	120 150 180	1.04 0.99 0.85	11.8 9.9 7.7	13.0 5.8	10
	120 150 180	1.24 1.18 1.07	22.1 20.7 18.2	23.5 15.0	15
4000	120 150 180	0.97 0.91 0.84	12.6 11.5 8.1	17.0 8.7	10
	120 150 180	1.11 1.05 0.97	24.6 22.5 18.8	28.5 17.0	15

Fig. 9 - Effect of opening pressure

To determine the usefulness of the duration signal, a series of tests were run. Fig. 11 shows the needle lift signal as well as the Timing Sensor output for a typical operating point. Laboratory tests showed that indeed the delivery can be expressed in terms of the duration signal. Fig. 12 plots the duration signal versus fuel delivery for a wide range of operating conditions. Output is very consistent.

Similar to tests conducted with the pintle nozzle, the opening pressure was changed and the effect noted. Unlike the indirect injector system, the changes in fuel delivery and duration were minimal. Only at cranking speeds was the change measureable. At operating conditions, the changes could not be separated from the run to run variation, and are not documented here.

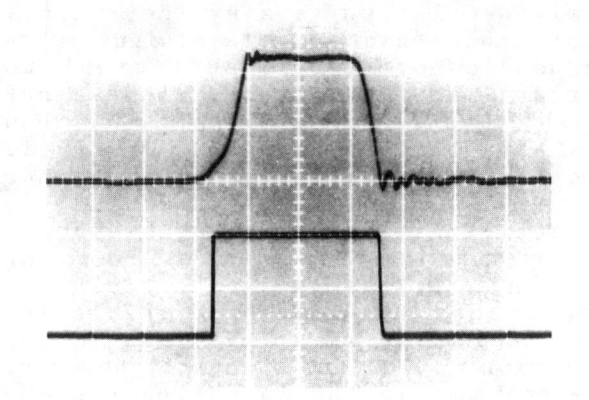

Fig. 11 - Needle lift and sensor signal

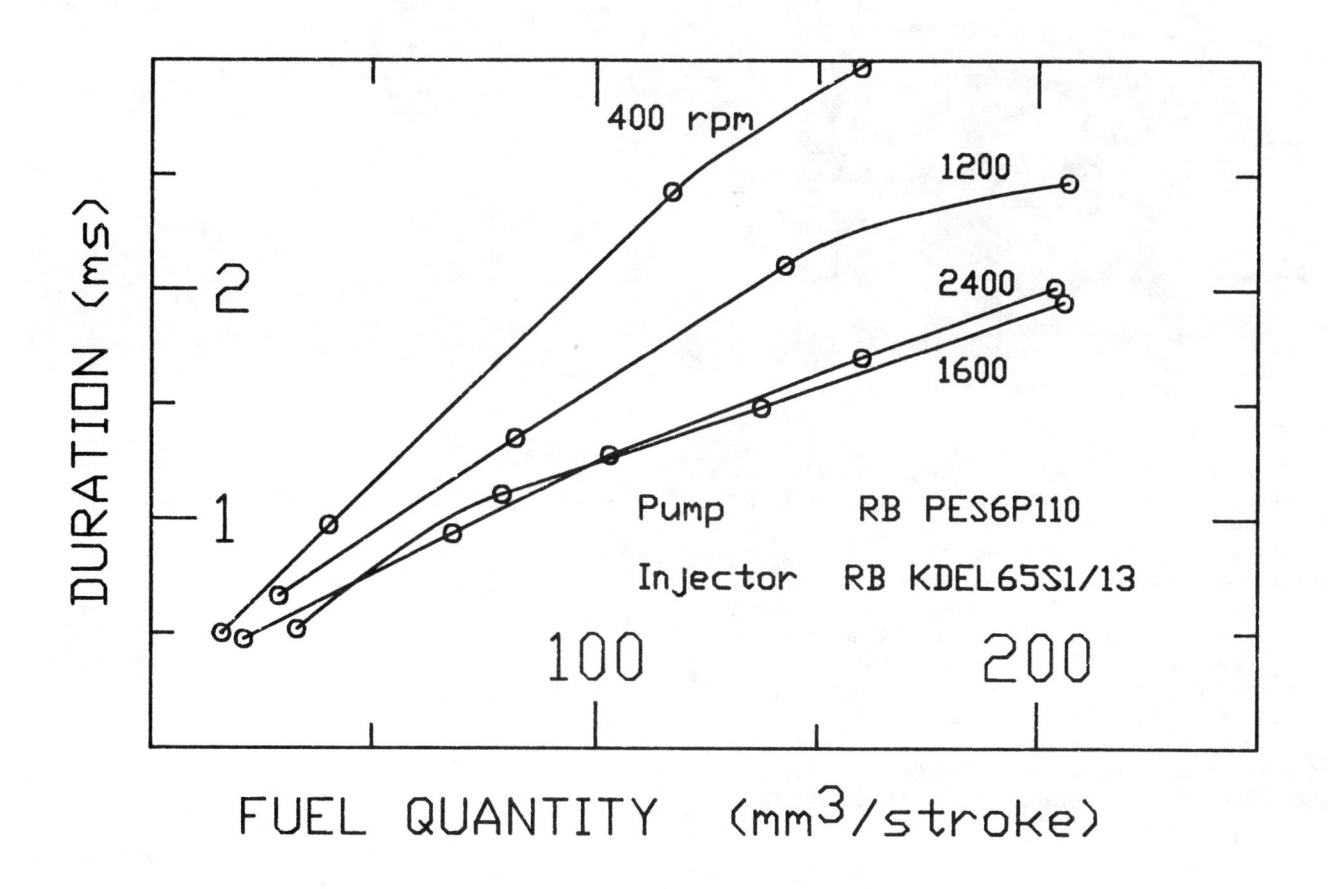

Fig. 12 - Timing sensor signal duration vs fuel quantity

Of concern, particularly with the direct injection systems, is needle bounce and secondary injections. Small signals generated by needle bounce are eliminated in two ways by the sensor. First, the sensor outputs a signal when the needle movement is sufficient to increase the flux density above a set threshold value; thus, signals less than this threshold are not acknowledged. Secondly, the nulling circuit responds to the small rapidly changing signals by effectively increasing the threshold. In the period after the bouncing perturbations, the nulling circuit re-establishes the threshold to the low level needed for beginning-of-injection detection. The sensor cannot, however, cure problems in the injection system. If a split injection or secondary injection occurs, it will be detected as would be a normal injection.

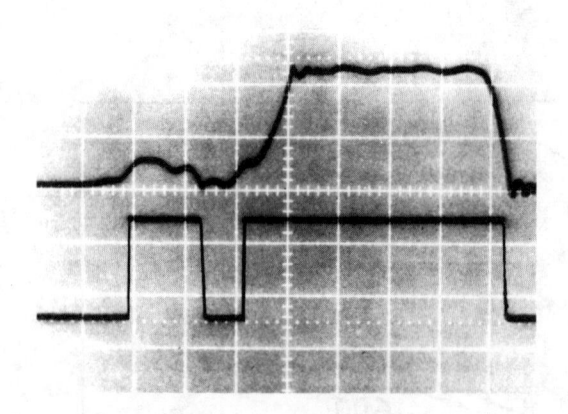

Fig. 13 - Split injection trace

Fig. 13 shows a split injection trace and the Timing Sensor output. Fuel quantity for the two part event must be specified as the sum of the duration signals. This may not be the case for a secondary injection event. This would have to be treated according to the use of the signal. For determining fuel consumption, both signals would have to be used. For a measure of engine load, the secondary signal would have to be omitted since it provides mostly smoke and little work.

CONCLUSION

From the data taken from a variety of indirect fuel injection systems as well as a direct injection system, it is obvious that the Wolff Controls Timing Sensor can be used to determine fuel quantity from its duration signal. It has been shown in a Driver Information System that the sensor can provide the fuel consumption information necessary to effect an average 10% savings on fuel (4). This will be the first production application of the sensor. The system will premiere on an European tractor this year.

The necessary modern control theory and microprocessor technology to implement an electronic diesel control system are already in existence. The actuators and sensors are under development. The Wolff Controls Timing Sensor has now proven that it is capable of providing the beginning-of-injection signal for timing control as well as a duration signal that can be related directly to fuel delivery for load control or fuel consumption analysis.

REFERENCES

1. "Diesel fuel injection system," Automotive Engineering, June 1983, p. 94.

2. R. Schwartz, "High-Pressure Injection Pumps with Electronic Control for Heavy-Duty Diesel Engines," SAE Paper No. 850170, 1980.

3. Gerhard Stumpp and Herman Kull, "Strategy for a Fail-Safe Electronic Diesel Control System for Passenger Cars," SAE Paper No. 830527, SAE special publication SP-540, 1983.

4. Eberhard Mausner, "Driver Information System for Tractors Electronic Components and Software Structure," ISATA Paper No. 810996, ISATA proceedings vol. 2, 1985.

5. J. Cambell, J. Scholl, F. Hibbler, S. Bagley, D. Leddy, D. Abata, and J. Johnson, "The Effect of Fuel Injection Rate and Timing on the Physical, Chemical, and Biological Character of Particulate Emissions from a Direct Injection Diesel," SAE Paper No. 810996, SAE special publication SP-495, 1981.

6. George D. Wolff and Michael S. Ziemacki, "Needle Lift Sensor Design Guide, Timing Sensor," Wolff Controls Corporation No. EI02-02/85, 1985.

7. George D. Wolff, "Smart injection sensor for electronic diesel controls," IAVD Paper No. C6,1, Int. J. of Vechicle Design, IAVD Congress on Vechicle Design and Components, 1985.

8. M. S. Ziemacki, "Diesel injector needle lift sensor," IMechE Conference publications 1985-12, Paper No. C212/85, 1985.

9. Milan Marcic and Zlatko Kovacic, "Computer Simulation of the Diesel Fuel Injection System," SAE Paper No. 851583, SAE special publication SP-630, 1985.

860144

Electronic Control of Diesel In-Line Injection Pump — Analysis and Design

Kazuro Nishizawa, Hiroshi Ishiwata, and Kenji Okamoto
Diesel Kiki Co., Ltd.

ABSTRACT

The Electronic Governor RED III for diesel in-line injection pumps was developed by Diesel Kiki, and introduced onto the market in 1983 as COPEC(Computed PE Control), which is a system for vehicles in combination with an electro-hydraulic timer. It has been well received by users because of its additional functions, i.e., auto-cruising etc, which enable the improvement of engine performance, drive-ability etc.

This report presents firstly an analysis of the stability of the engine-governor system, and secondly an outline of the COPEC system from the point of view of engine stability, reliability, etc.

Additionaly a growing need exists for electronic control of various functions, such as generator and construction machine operation. In this paper is also described how the control unit (C/U) answers these various needs, as the second stage of the first generation electronic control system.

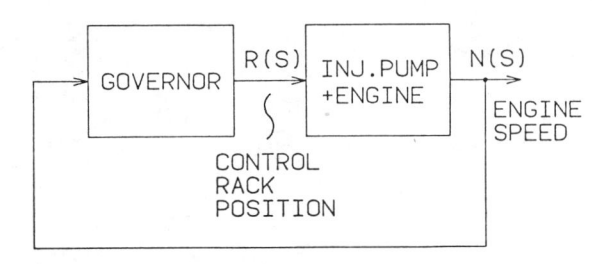

Fig.1 Outline of governing system

---

* Numbers in parentheres designate references at the old paper.

STABILITY ANALYSIS OF ENGINE-GOVERNOR SYSTEM

Fig.1 shows a block diagram of the engine-governor system. Whether a hunting will occur or not, is estimated by finding the transfer function of each element and applying the stability criterion to the total system. We would like to pursue the argument using the generator-system, which is one of the most difficult system for obtaining stability.

EXPERIMENTAL APPROACH -- By measuring the frequency characteristics of the engine (Rack position → Speed) and governor (Speed → Rack position) with the FFT analyzer, and adding these two bode diagrams in the frequency domain, stability criterion can be found by estimating the phase and gain margin. Fig.2 shows an example.

Open loop gain is obtained by adding the measured governor gain and the engine gain. Open loop phase is simultaneously obtained in the same way. If the open loop gain is positive at the frequency where open loop phase = -180°, the system is stable and this positive gain value is determined as the gain margin. In a similar way the phase margin is determined by estimating the phase disparity to the -180° line at the frequency where open loop gain=0.

Fig.2 shows the experimental result using 2 kinds of flywheel inertia. In both cases, the system is stable but it can be also seen that the gain and phase margin are larger when the inertia is large. And in addition, when we try to obtain decreased droop with a mechanical governor, phase and gain margin will be drastically decreased, because of the higher open loop gain and the larger phase lag of the governor, caused by a smaller spring constant. In the case of the mechanical governor, a Change of statical characteristics causes a simultaneous change in dynamic response.

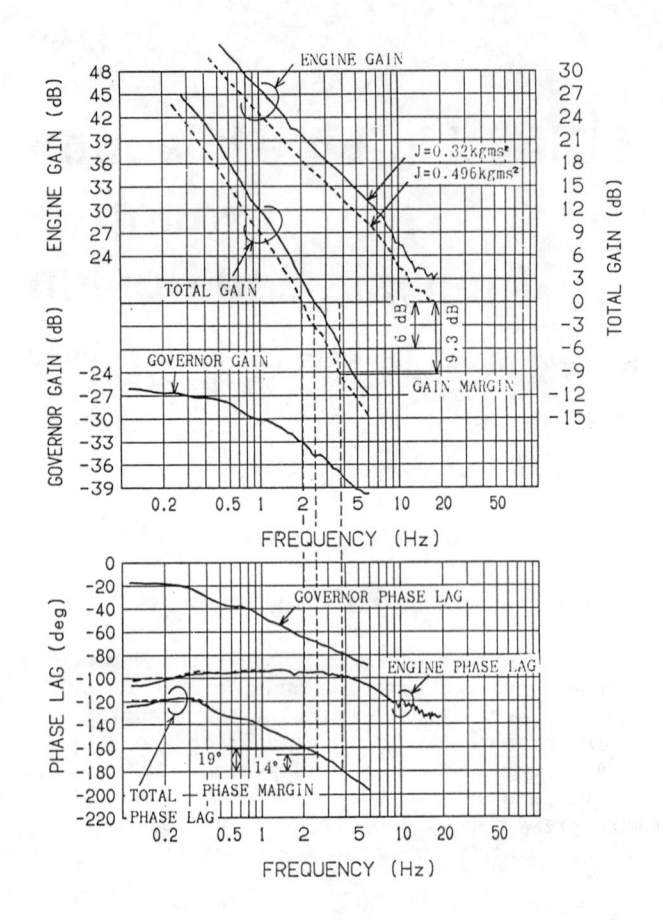

Fig.2 Bode diagram of engine control system

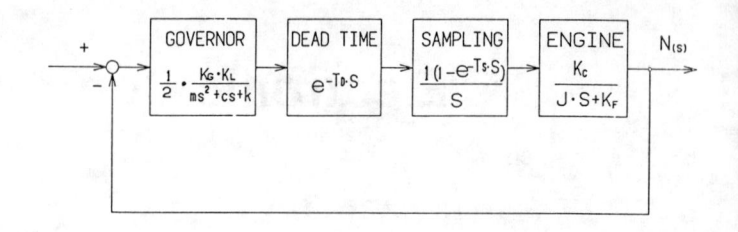

Fig.3 Modelling of governing system

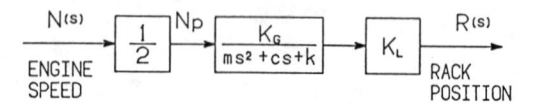

Fig. 4 Transfer function of mechanical governor

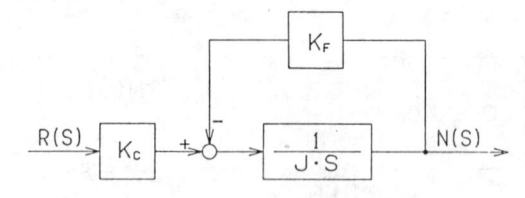

$$G_{g(s)} = \frac{R_{(s)}}{N_{(s)}} = \frac{1}{2} \cdot \frac{K_G}{ms^2 + cs + k} \cdot K_L$$

m : MOVING MASS
c : VISCOSITY COEFFICIENT
k : SPRING CONSTANT
K_G : GOVERNOR GAIN
K_L : LEVER RATIO

Fig.5 Transfer function of injection pump & engine

$$G_{g(s)} = \frac{N_{(s)}}{R_{(s)}} = \frac{K_c}{J \cdot S + K_F}$$

J : MOMENT OF INERTIA
K_c : Δ TORQUE/Δ RACK
K_F : FACTORS VARYING WITH N_E (INJ.QUANTITY, TORQUE etc.)

ROOT LOCUS METHOD -- By finding the characteristic equation of the total engine-governor system, and solving it using the Newton-method, root locus is aquired. Fig.3 shows the modelled block diagram, Each element will be described as follows.

(Governor).......Estimated as a spring-mass system.
(Dead time)......Summerised dead time in the total loop will be treated using Padé approximation.
(Sampling lag)...Sampling lag exists in the engine-governor system, because the transfer from rack position to torque is an intermittant phenomenon which occurs at each combustion. [1]*[2]*
(Engine).........Estimated as an inertia system, which translates the rack position to engine speed through combustion. Kf includes all factors, which vary with engine speed, such as friction, injection quantity etc.   -- Fig.5

The open loop transfer function will be described as

$$Go(s) = \frac{1}{2} \cdot \frac{K_G \cdot K_L}{ms^2+cs+k} \cdot \frac{-\frac{T_d}{2} \cdot s+1}{\frac{T_d}{2} \cdot s+1} \cdot \frac{1}{s} \cdot (1-\frac{-\frac{T_s}{2} \cdot s+1}{\frac{T_s}{2} \cdot s+1}) \cdot \frac{K_C}{J \cdot s+K_F} \quad (1)$$

and the characteristic equation ;

$$1+Go(s)=0$$

$$1+\frac{1}{2} \cdot \frac{K_G \cdot K_L}{ms^2+cs+k} \cdot \frac{-\frac{T_d}{2} \cdot s+1}{\frac{T_d}{2} \cdot s+1} \cdot \frac{1}{s} \cdot (1-\frac{-\frac{T_s}{2} \cdot s+1}{\frac{T_s}{2} \cdot s+1}) \cdot \frac{K_C}{J \cdot s+K_F} = 0 \quad (2)$$

By substituting concrete values to equation (2), and solving it using the Newton-method, the root locus is obtained on the complex plane. .....Fig.6.

Because the dominant root, which is the nearest to the imaginary axis, will dominate the dynamic behavior of the system, it is sufficient only to investigate the dominant root locus. Fig.7 shows an example obtained by varying the moment of inertia of the engine and the speed droop. It is seen here that larger inertia enables better engine stability.

Fig.8 shows some results where we have experienced the hunting criterion. Correlation is found between the real part values of the dominant root and hunting results. In the case of the droop under 4 %, it is considered difficult to solve the hunting problem with the current engine - mechanical governor system.

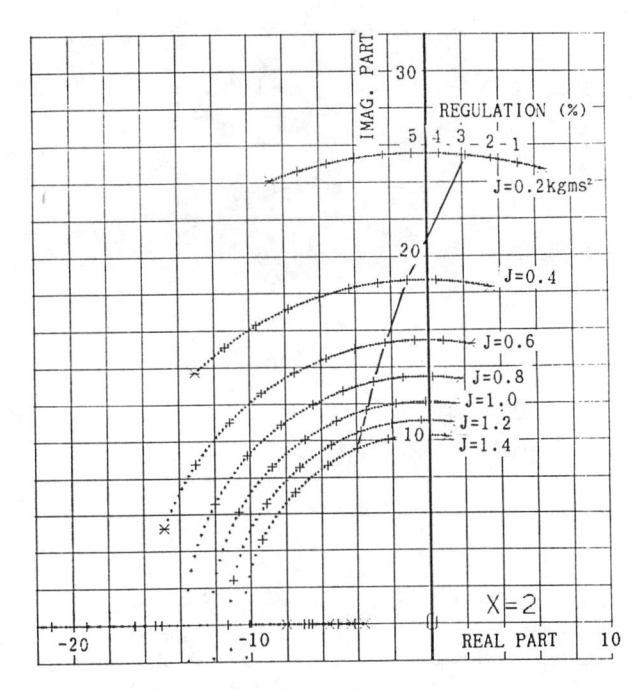

Fig.7 Dominant root locus by varying the moment of inertia and speed droop

⊙ : UNDER MASS PRODUCTION
△ : CRITICAL
X : HUNTING

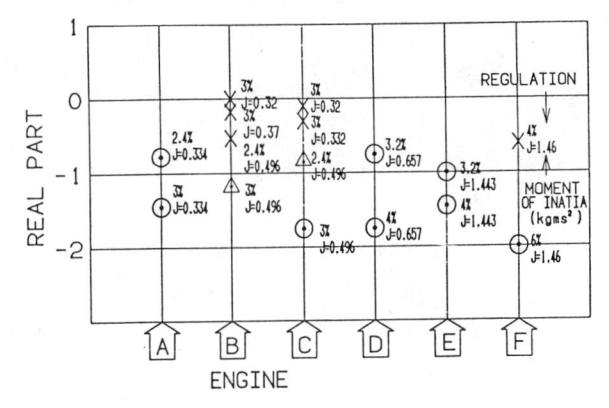

Fig.8 Engine hunting and real part of the dominant root

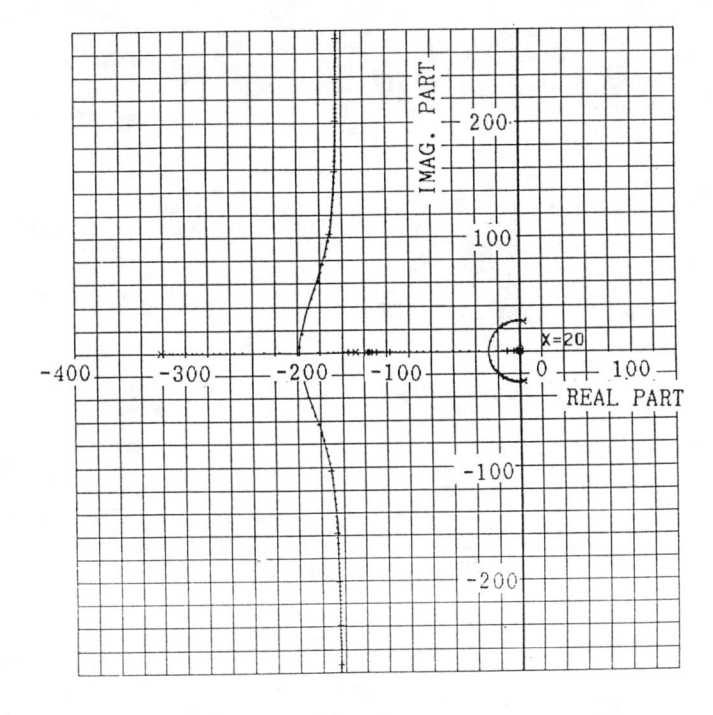

Fig.6 Root locus of engine-governor system

The next theme concerns the improvement of the stability by applying the larger control force of the governor flyweight. Fig.9 shows the relation between the real part of the root and governor control force, where determining current force = 100. The larger flyweight is less effective when the droop is 3 %. It suggests that the droop limit is inevitable in the current mechanical governor-engine system.

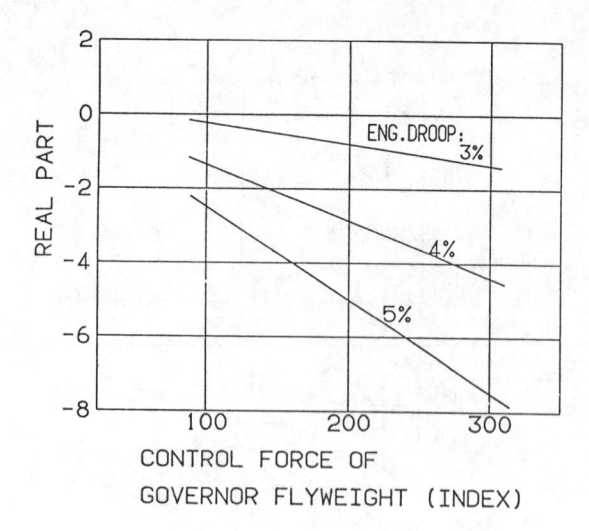

Fig.9 Effect of control force

On the other hand, in the engine-governor system, there are many elements that are considered to have dead time. Fig.10 shows the dominant root locus obtained by varying the dead time and the droop. As an example it can be said that the dead time of 10 msec. corresponds to the droop change of 1 % in the region of droop=4 %. Dead time causes adverse effects on engine stability. This is an important point when the digital control system, the main theme of this report, is to be investigated.

ENGINE SPEED CONTROL WITH ELECTRONIC GOVERNOR

Taking the above into consideration, we will present how Diesel Kiki RED III was designed.

The electronic governor will be separated into two blocks(Fig.11), the servo system and the calculation system.

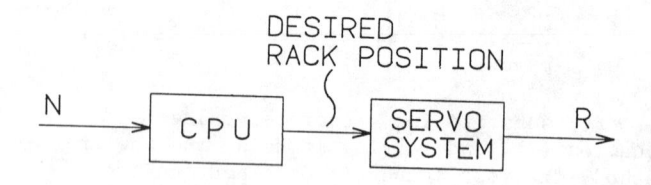

Fig.11 General block diagram of electronic governor

SERVO SYSTEM --
Actuator -- Fig.12 shows the schematic view of RED III. A magnetic field in a radial direction is formed at the cylindrical air gap, and the moving coil is placed in this gap area. Force is obtained by applying the current, and the direction of the force can be changed by changing the direction of the current. This is the so-called 'LINEAR DC MOTOR'.

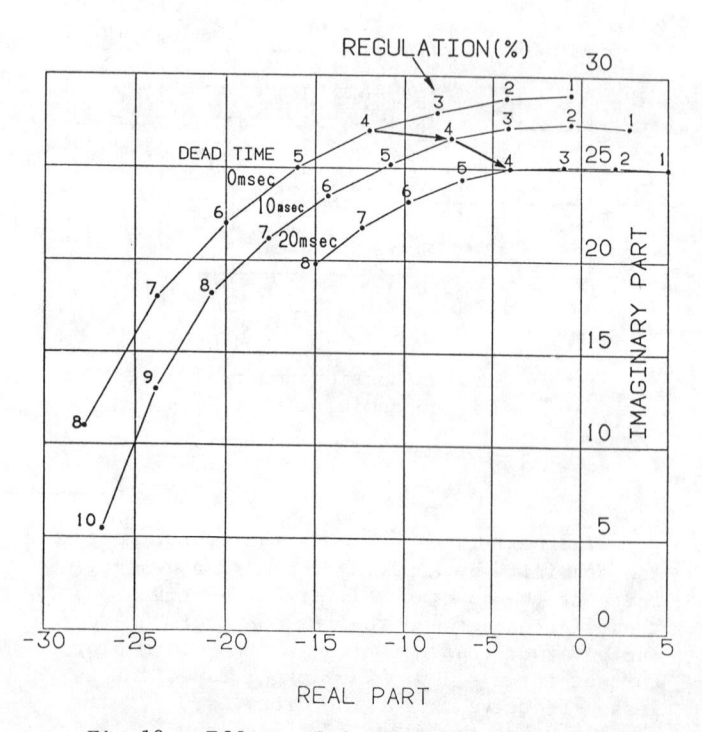

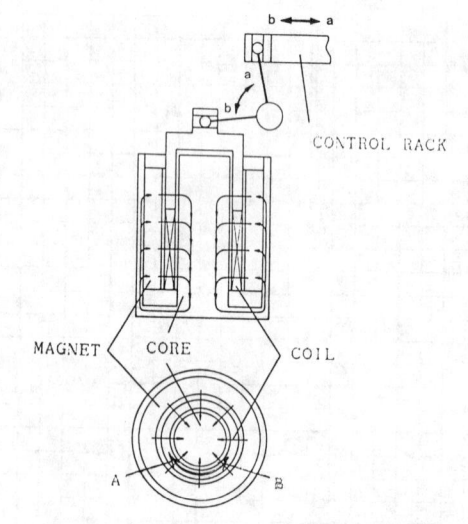

Fig.10 Effect of dead time

CURRENT	FORCE	ACTION	INJECTION QUANTITY
A	UP	a	INCREASE
B	DOWN	b	DECREASE

Fig.12 REDIII actuator

Electronic governor must satisfy the following conditions.

(For cold starting)
- sufficient force, taking into account loss through spring force
- less friction resistance
- less viscous resistance

(For good response)
- larger dynamic range of the force, including minus direction
- smaller moving mass
- less friction resistance
- less viscous resistance

(For reliability)
- less joule heat

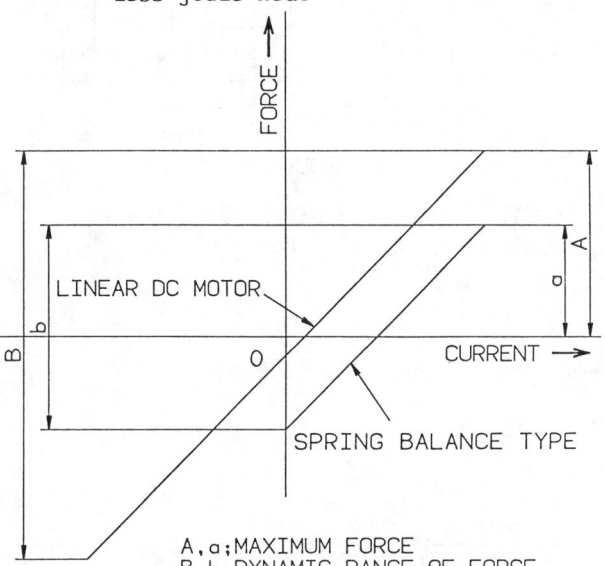

Fig.13 Comparison of force:
..Linear DC Motor & Spring balance type

2 types of actuator can be considered for diesel injection pumps ; one is the linear DC motor (current direction switching type) and the other is the spring balance type. We have chosen the linear DC motor, taking the above-mentioned conditions into consideration. In view of force, the linear DC motor can be seen to have an advantage.-- Fig.13. In this figure, it is assumed that the actuators have the same force when the current is the same. The spring balance type seems to be inferior from the point of view of the maximum force, because the spring force behaves as a resistant force. In the case of the linear DC motor, a wide dynamic range will be obtained because of the existance of the minus direction force.

In addition, in order to obtain the same force, the linear DC motor requires less line of magnetic force (current x number of turns) than the spring balance type because of the existance of the permanent magnet.

Construction of the actuator ass'y -- An eddy-current type rack sensor is used as a rack position sensor. The actuator ass'y consists of 3 parts, the housing, the middle plate ass'y, and the cover.(Fig.14) This enables the concentration of the wiring to the middle plate ass'y. The actuator contains no oil, but the oil swash from the pump chamber is available for linkage lubrication.

Tab. 1 shows the combination of pump and actuator. The sub-coil enables less starting time because of its additional magnet field. (Fig.15)

CYL. NO. PUMP TYPE	4	6	8	10	12	6 H
A	S	S	S	/	/	/
P	/	S	C	C	C	C
ZW	/	C	C	/	/	/

H : HORISONTAL MOUNTING
S : STANDARD TYPE
C : WITH SUB-COIL

Tab.1 Applicable pumps for REDIII

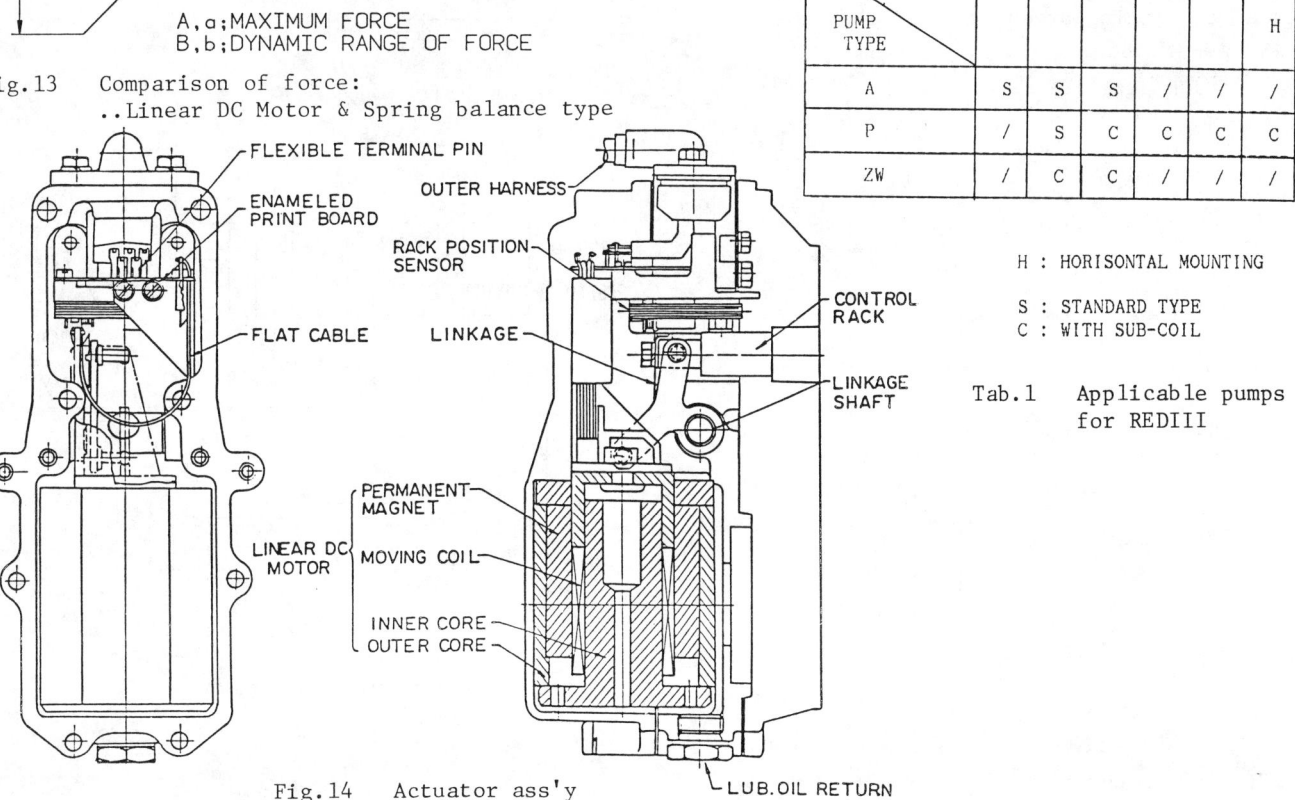

Fig.14 Actuator ass'y

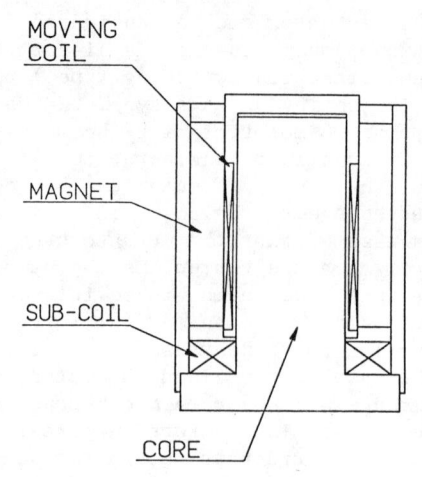

Fig. 15 Sub-coil added to actuator

Performance of the servo system -- Fig. 16 shows the measured force of REDIII. At cold starting, by using the pre-acting mode, it is possible to aquire the staring rack position down to a temperature of -30°.

Fig. 17 shows the frequency characteristic of the servo system, which consists of the driving circuit, the actuator, and the rack sensor.(Fig. 18) The measured characteristics shows its advantage over the mechanical governor.

When we consider the dynamic response of the governor, response to small amplitude input is very important in view of the engine hunting, vehicle surge, and fine correction of the rack position, which will be mentioned later. Fig. 19 shows the advantage of the nonlinear element installed in the servo circuit in order to obtain larger gain at the small amplitude.

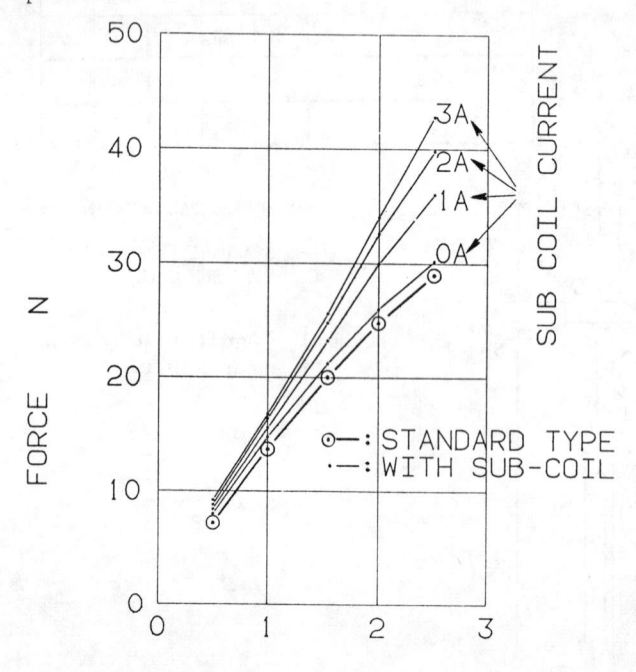

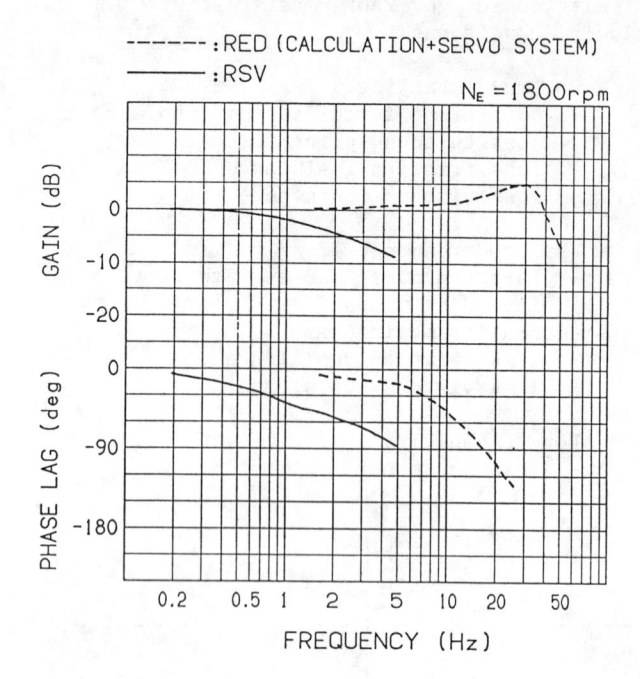

Fig. 17 Frequency characteristics of servo system

Fig. 16 Actuator force of REDIII

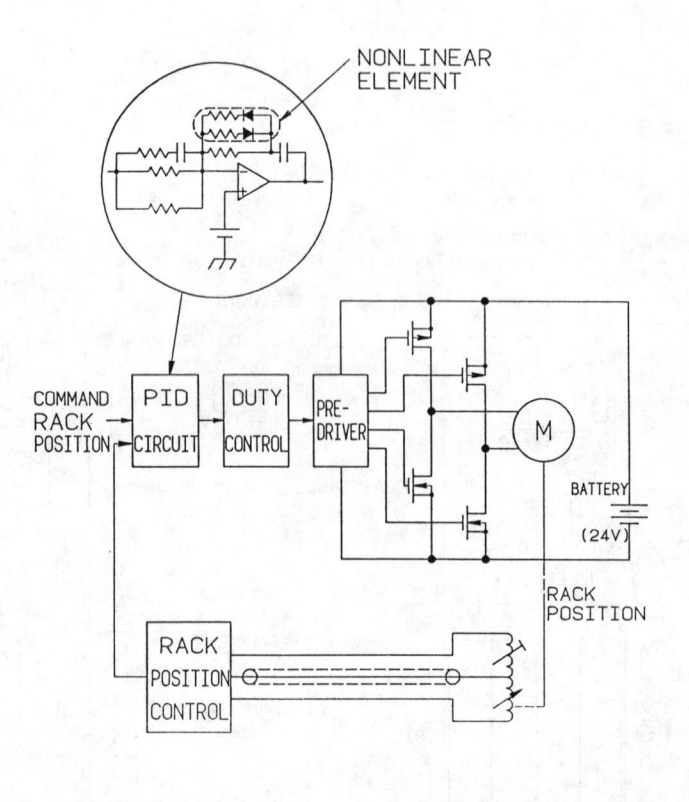

Fig. 18 Rack servo system

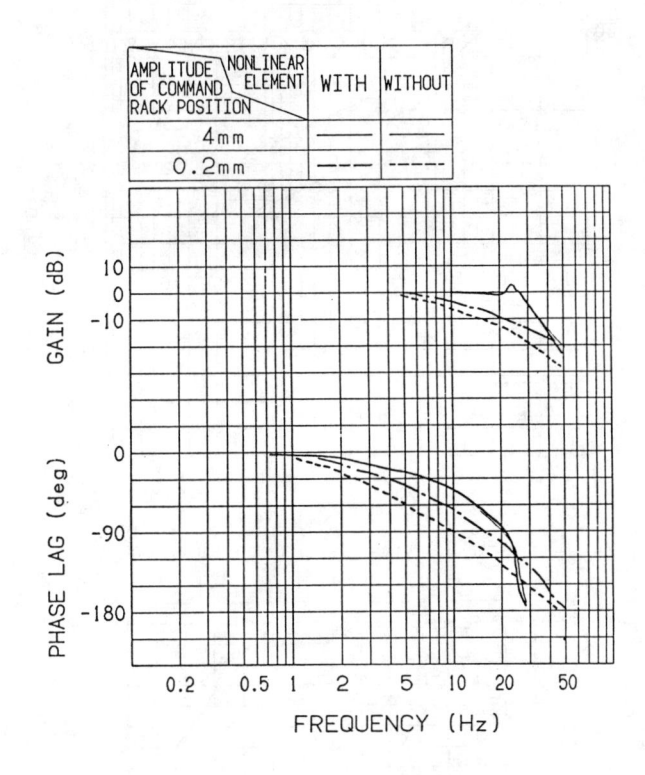

Fig.19 Effect of non-linear element

Tab.2 Calculation method of desired rack position

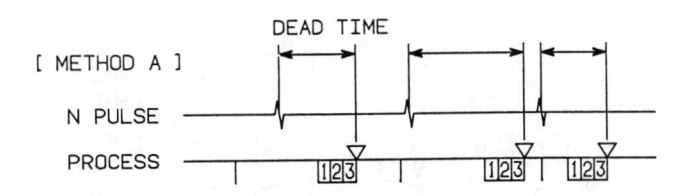

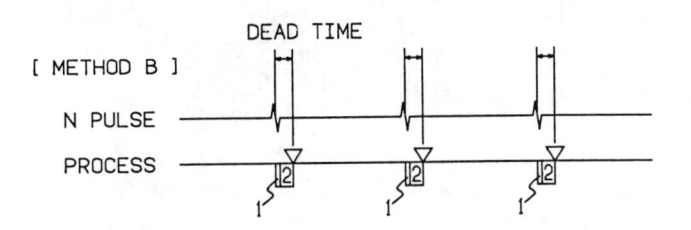

	METHOD A	METHOD B
TIMING OF GOVERNING CALCULATION	CONSTANT INTERVAL	SYNCHRONIZED WITH N PULSE
ENGINE SPEED CALCULATION	DIVIDING CONSTANT BY N PULSE INTERVAL	TABLE LOOK UP METHOD
Q → R CONVERSION	CALCULATE TARGET FUEL QUANTITY AND CONVERT IT TO RACK POSITION	CALCULATE RACK POSITION DIRECTLY

1: ENGINE SPEED CALCULATION
2: GOVERNING CALCULATION
3: Q → R CONVERSION

Fig.20 Timing chart of electronic governor

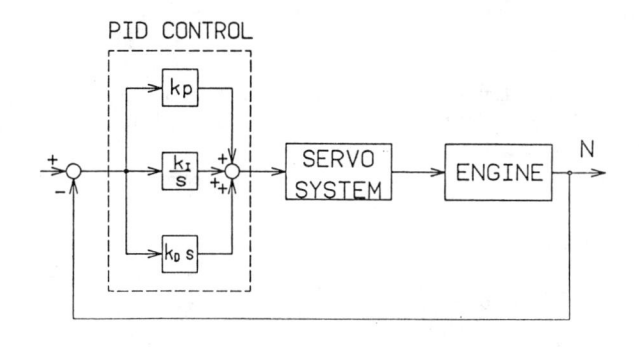

Fig.21 Block diagram of isochronous control

CALCULATION SYSTEM --

CPU -- INTEL 8031 is used with 12MHz clocking. This is a control-oriented CPU without program memory, which also has a multiplication command.

Basic software -- As to calculating the desired rack position, we have used the method A of Tab. 2, but in order to obtain a better response, i.e. to reduce the dead time of the system we have changed to B. Fig.20 shows the advantage concerning dead time.

Merits of digital control --

(1) Generator set : as mentioned before, the droop limit is approximately 3 % in the case of the mechanical governor-engine system. But with the electronic governor, 0 % droop is possible using PID control. Fig.21 shows this system block-diagram. Kp,Ki,Kd are adjustable from outside of each generator, in order to obtain the best stability and momentary droop. Fig.22 shows the typical bode diagram of open loop PID control.

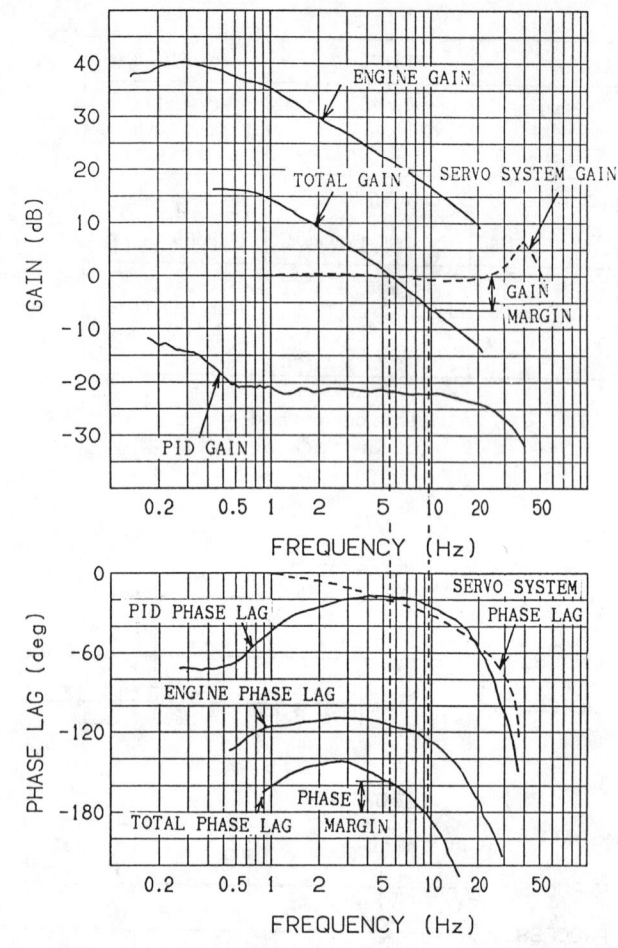

Fig.22 Bode diagram of isochronous control

(2) Compensation elements (Fig.23), such as phase lag compensation or phase lead compensation are available. Fig.24 shows an example, in which the phase lag compensation is applied. The merit of compensation is apparent here.

(3) The static and dynamic characteristics can be obtained individually, and flexible control is possible by inputting the constants into characteristics table. In the case of the mechanical governor, dynamic response is subordinated to the static characteristic [3]*.
--- Fig.25

(4) Individual cylinder control at idling is available. This enables the reduction of vibration of the vehicle cabin.
--- Fig.26

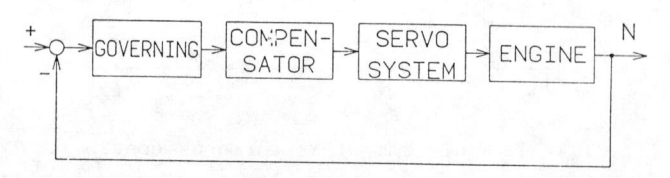

Fig.23 Block diagram of electronic governor with compensation

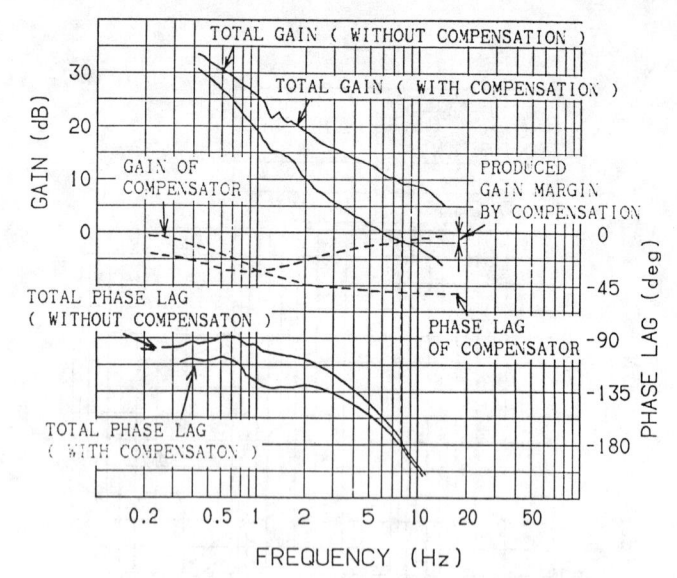

Fig.24 Effect of phase lag compensation

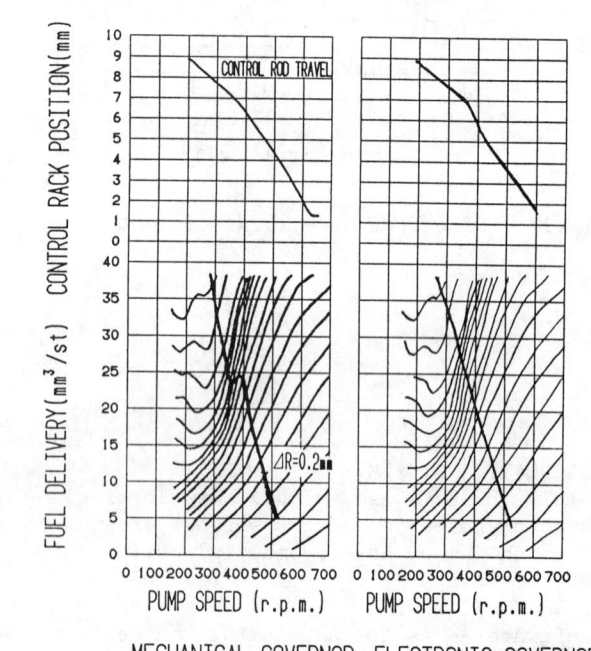

Fig.25 Effect of flexible N-R characteristics

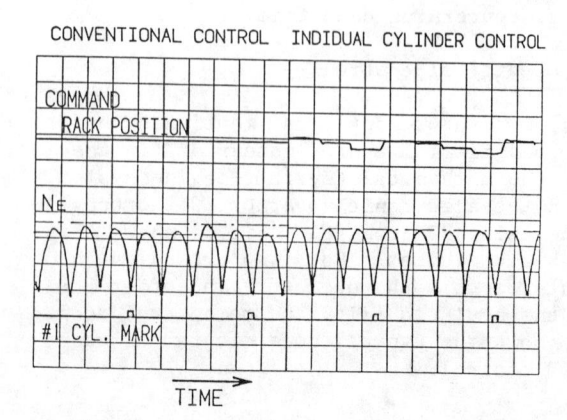

Fig.26 Individual cylinder control

COPEC

GENERAL INFORMATION OF COPEC -- The system, which is combined with the electro-hydraulic timer, is called COPEC (Computed PE Control). A typical block diagram and the construction are shown in Fig.27 and Tab. 3.

As COPEC is applied to vehicle, additional functions such as idling control, auto-cruising etc. are necessary. Timing can be controlled as shown in Fig. 29. This will be explained in detail in Follwing [4]*.

1. INJECTION PUMP ASSEMBRY
    INJECTION PUMP PROPER
    GOVERNOR ACTUATOR
    TIMER ACTUATOR(with timing sensor)
2. CONTROL UNIT
3. TIMING CONTROL VALVE
4. WIRE HARNESS
5. SENSORS
    TDC(N) SENSOR
    ACCEL POSITION SENSOR
    COOLANT TEMPERATURE SENSOR
    BOOST PRESSURE SENSOR
6. SWITCHES
    START SW.
    HEAT SW.
    CRUISE MAIN SW.
    CRUISE SET SW.
    CRUISE RELEASE SW.
    CRUISE RESUME SW.
    DIAGNOSIS SW.
7. LAMPS
    DIAGNOSIS LAMP, CRUISE LAMP
    ECONOMY LAMP

Tab.3 Components of COPEC

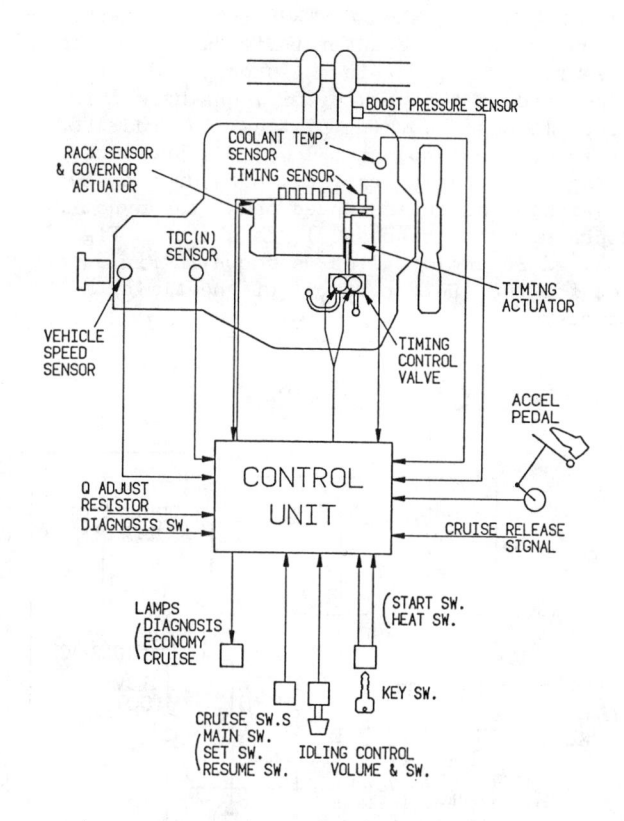

Fig.27 COPEC system diagram

R :CONTROL RACK POSITION
Tc:COOLANT TEMPERATURE
Ne:ENGINE SPEED
Ap:ACCELERATOR PEDAL POSITION
PB:BOOST PRESSURE

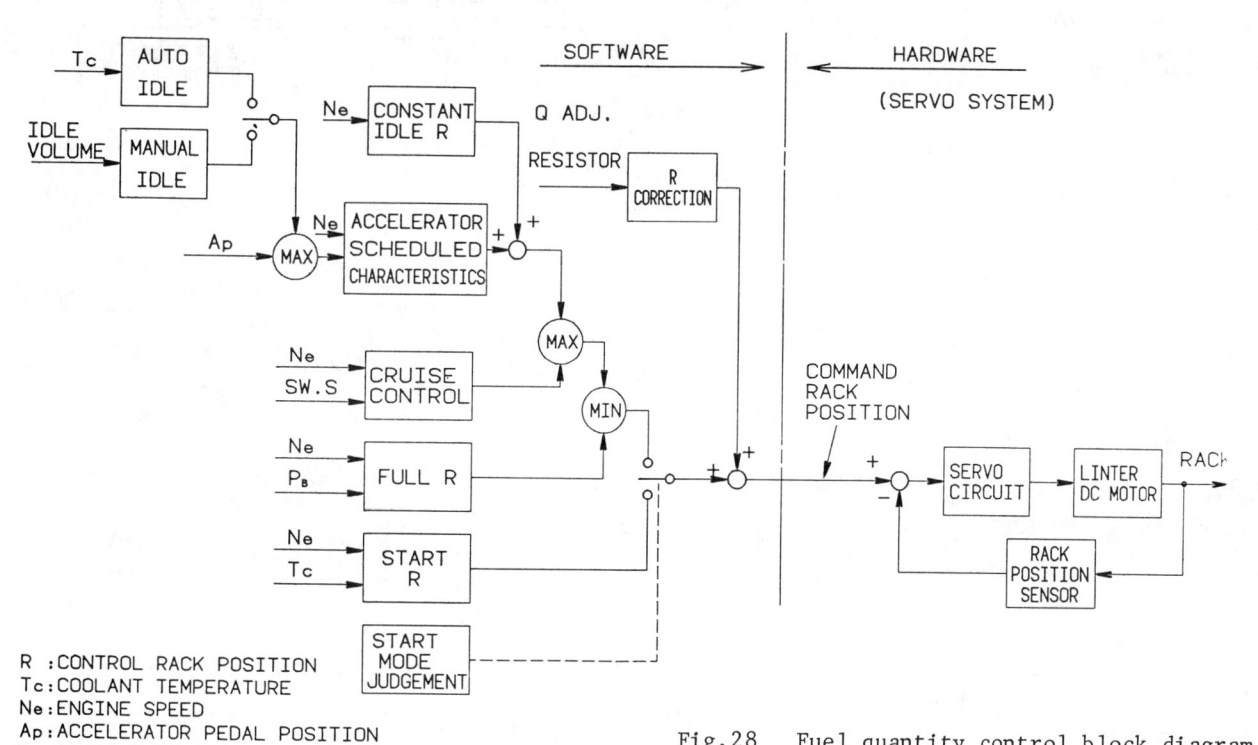

Fig.28 Fuel quantity control block diagram

INJECTION TIMING CONTROL -- The electro-hydraulic timing actuator (Fig.30) consists of eccentric cams, working pistons, and other parts, and as the working oil, engine oil is used. It has 4 working pistons in order to obtain good control at a lower oil pressure. Timing characteristics (Oil pressure=constant) against the engine speed, and the dynamic response (step response) are shown in Fig.31 & 32. 2 Solenoid valves as shown in Fig.33 are used for the 'Duty control' of the timing actuator.

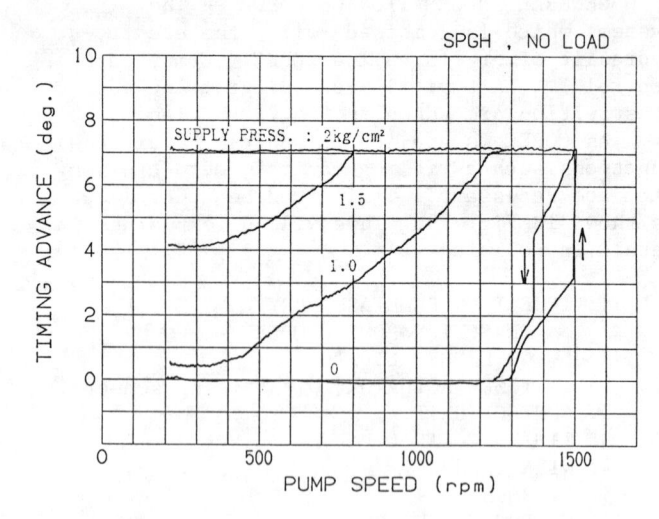

Fig.31 Timing advance characteristics for working oil pressure

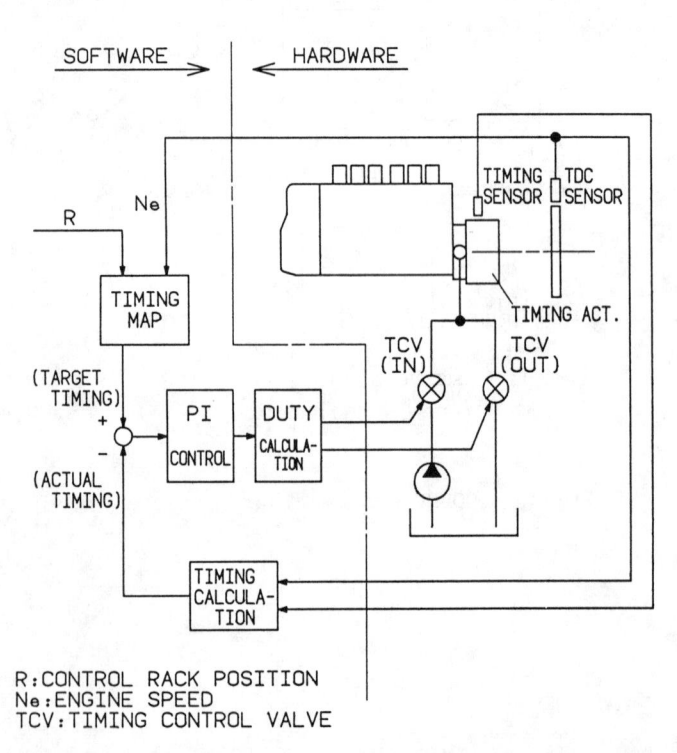

Fig.29 Block diagram of injection timing control

R:CONTROL RACK POSITION
Ne:ENGINE SPEED
TCV:TIMING CONTROL VALVE

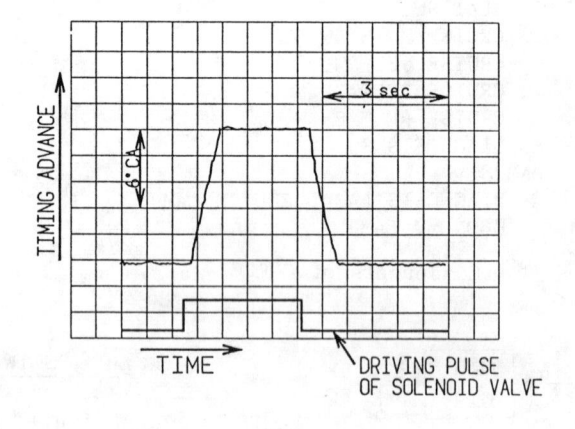

Fig. 32 Dynamic response of the timing control

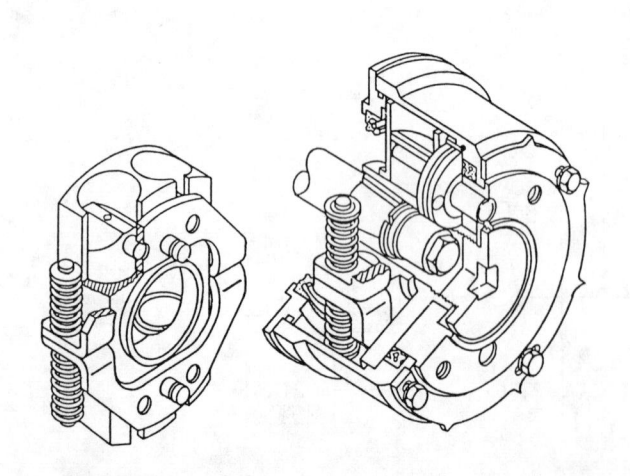

Fig.30 Hydraulic timing device

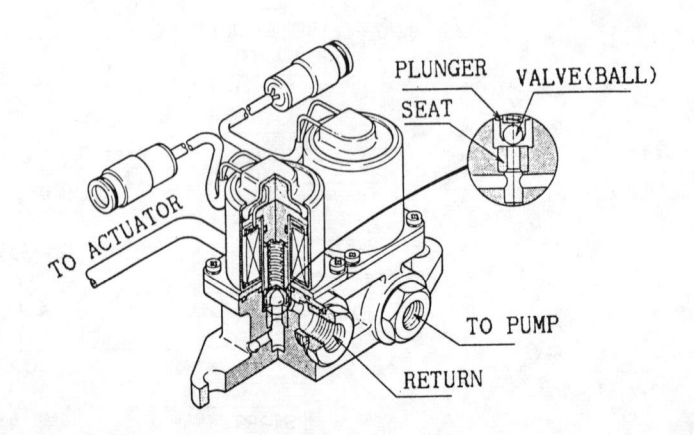

Fig.33 Solenoid valve

SERIES II CONTROL UNIT

BASIC SYSTEM -- Because of the flexibility of diesel electronic control, there now exists a growing need for its application. We have been investigating how to respond to these various needs, and have developed a new series of the control unit, named Series II. Its functions and the basic hardware (case and printboard) are shown in Tab.4. The following points have been considered.

- The basic system can be constructed with main printboard only, on which are the basic parts, in order to realize the indispensable functions of the governor.
- In the case of the basic system, connectors are installed on the main printboard. (case Fig.: A of Tab. 4)
- The interface for serial communication is on the main printboard, because its function has a close relation to the CPU core construction.
- The interface and connector for the expansion system are on the 2nd printboard. ...parts on the main printboard are always the same.
- A common frame is used for all systems.
- A high roof cover is prepared for the expansion system in which the parts are on the upper side of the 2nd printboard. (case Fig.: B of Tab.4)

The hardware of the basic system is shown in Fig.34 & 35. A number of hybrid ICs are used.

For software development, an operation system for 8051 has been prepared, where new functions are easily added to the system.

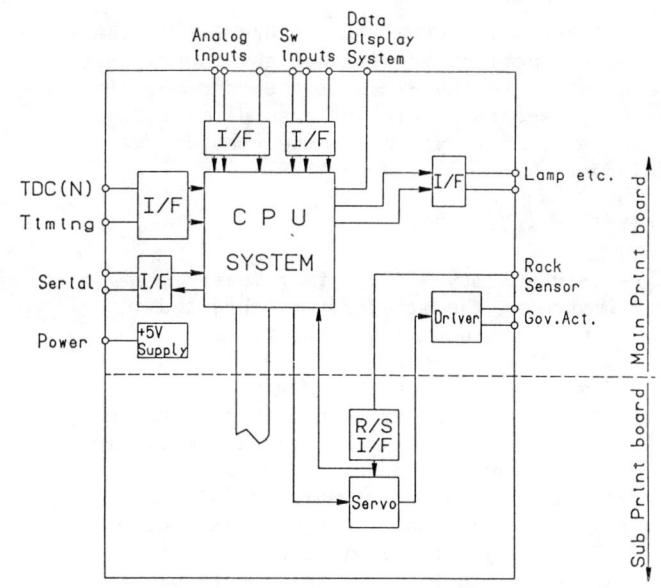

Fig.34　Series II control unit basic system

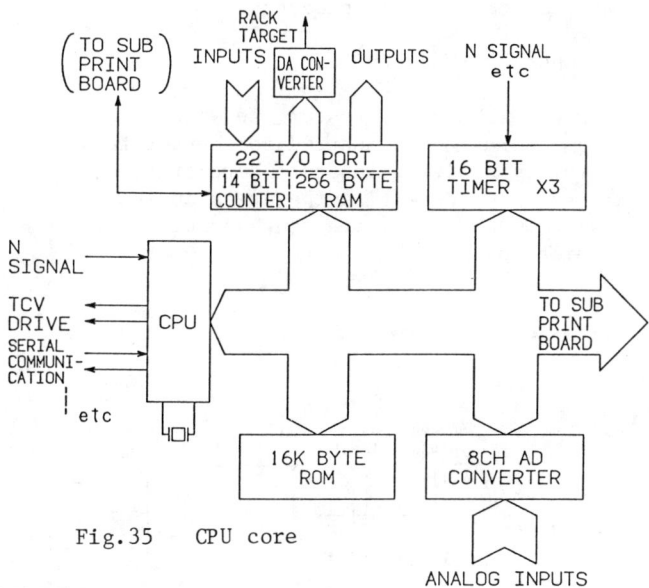

Fig.35　CPU core

Tab.4　Control unit needs & applications

APPLICATION OF SERIES II C/U -- These are various functions required of governors. Responding to the needs of the corresponding engine, especially in the case of electronic governors, many systems can be considered, as follows.

(1) Trucks(...COPEC)

- Driveability will be improved by matching the accelerator scheduled injection quantity characteristics.
- Variable timer function enables the improvement of the accuracy.
- Confortable driving through auto cruising.
- System construction is already shown in Fig.28 and the hardware in Fig.36.

Control block diagram of auto cruising is shown in Fig.37. Electronic transmission control is becoming a major consideration for vehicle development, and it is profitable to use the data jointly. For this purpose, serial communication has been prepared.

(2) Construction machines

Fig.38 shows the system, in which the optimum operation is assured for better fuel economy, by applying the auto-deceleration function, i.e. detecting the exess engine rpm. Serial communication with the hydrqulic system is available.

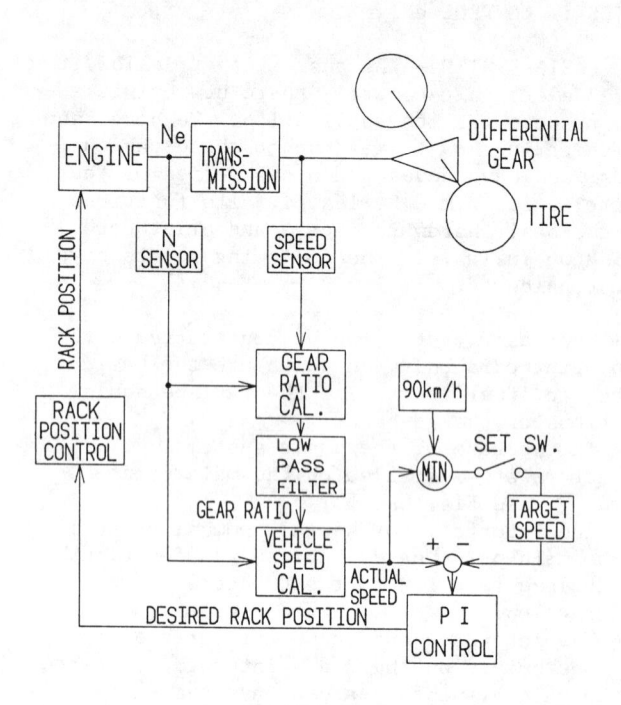

CAL.: CALCULATION

Fig.37 Bolck diagram of auto-cruising control

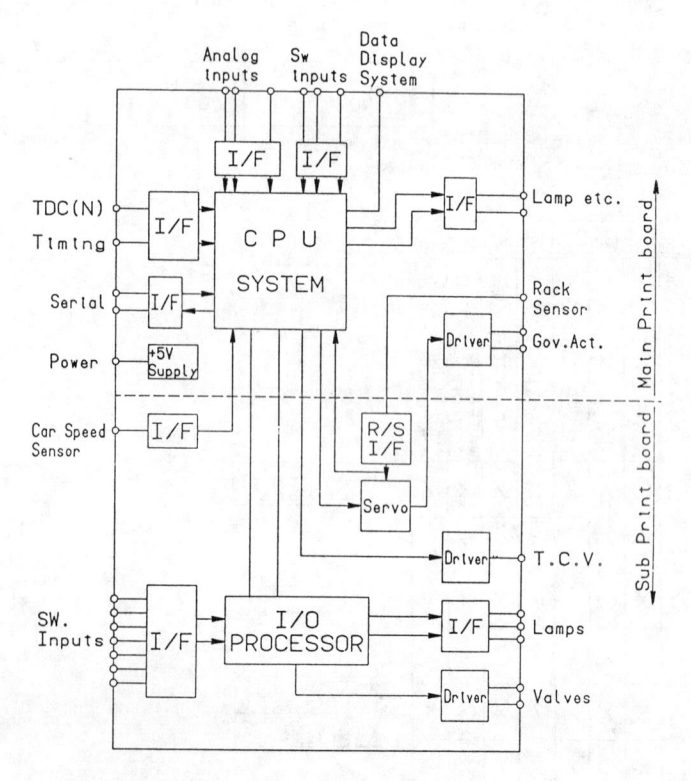

Fig.36 Control unit for vehicles

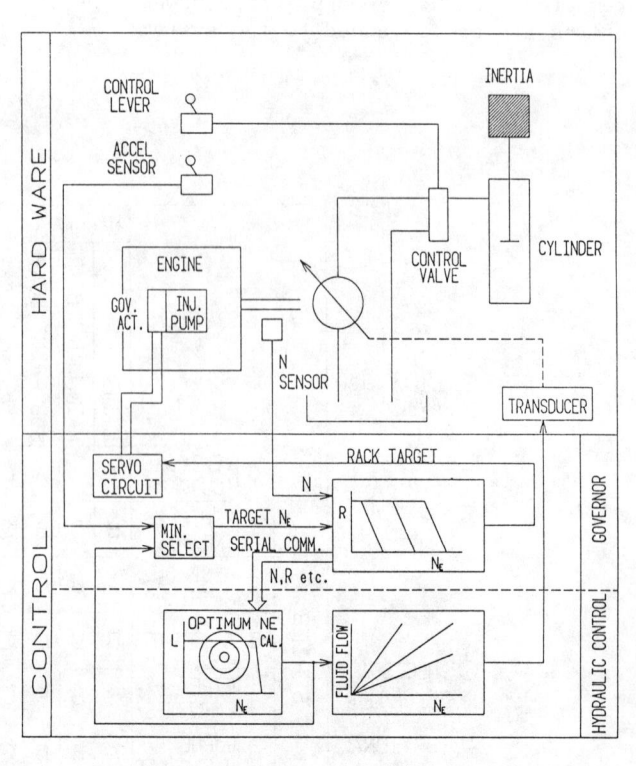

Fig.38 Construction machine control

(3) Generators

By applying PID control, 0 % droop (isochronous) is possible, and it is also possible to minimize the momentary droop through load current feed back. This C/U has adjustable volumes for PID constants, which enable the adjustment of momentary droop from outside, corresponding to the inertia of the generator system. (Fig.39)

(4) Fire engines

The system for fire engines will have a water pressure control function. Hardware will be as shown in Fig.40 because the interface for the water pressure sensor is needed.

(5) Heavy duty V-engines

In the case of heavy duty V-engines, there are cases where 2 pumps are mounted on one engine. In these cases, governor actuator for each pump is available, and both actuators are controlled by one control unit. Fig.41 shows the block diagram.

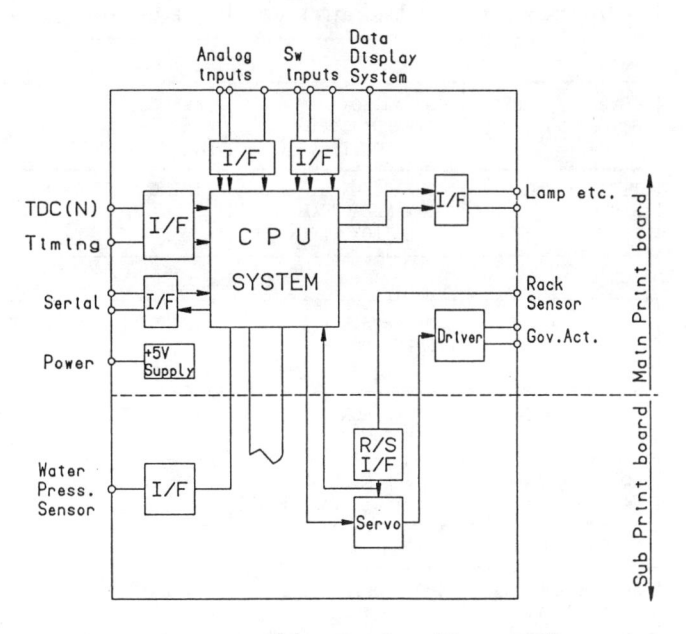

Fig.40 Control unit for fire engines

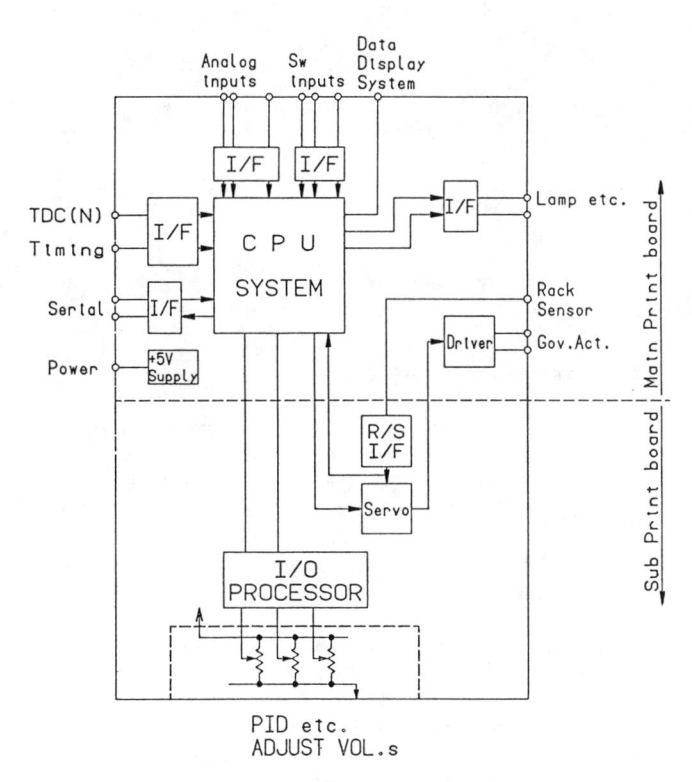

Fig.39 Control unit for generators

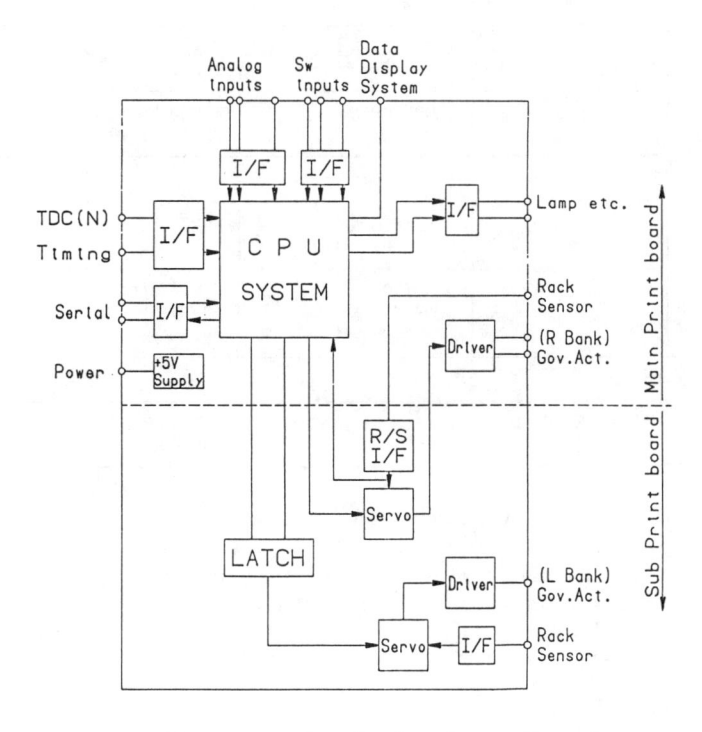

Fig.41 Control unit for large V-engines

RELIABILITY -- Although each component of the system has already been evaluated by various test modes, the system must have a back up system for emergencies. Tab.5 shows the on-board diagnosis, its sensing method, and the treatment after sensing.

103

COMPONENTS	FAILURE DETECTING	BACK UP
< CONTROL SYSTEM >		
GOVERNOR SERVO SYSTEM	COMPARISON COMMAND/ACTUAL RACK POSITION	FUEL CUT *
TIMER SERVO SYSTEM	COMPARISON COMMAND/ACTUAL INJECTION TIMING	TIMING CONTROL STOP
< PICK UP >		
TDC SENSOR	COMPARISON WITH TIMING PULSE	TIMING CONTROL STOP USING RPM CALCULATED BY TIMING PULSE
TIMING SENSOR	COMPARISON WITH TDC PULSE	TIMING CONTROL STOP
BOTH TDC AND TIMING SENSOR	PULSE INTERVAL (ONLY AT STARTER ON)	FUEL CUT
< ANALOG INPUTS >		
ACCELERATOR PEDAL POSITION	SHORT/OPEN	GOVERNING AT FIXED RPM
COOLANT TEMP.	SHORT/OPEN	FIXED TO 80 DATA
IDLING VOLUME	SHORT/OPEN	FIXED TO AUTO IDLING MODE
FUEL QUANTITY ADJUST RESISTOR	SHORT/OPEN	FIXED TO STADARD Q DATA
BOOST PRESSURE	SHORT/OPEN	FIXED TO LOW PRESSURE DATA
< SYSTEM >		
MEMORY (RAM)	COMPARISON WRITE/READ DATA	FUEL CUT
AD CONVERTER	SENSING COMPLETION	ALL ANALOG DATA ARE FIXED TO BACK UP MODE
< OTHER >		
VEHICLE SPEED	PULSE INTERVAL	RELEASE CRUISE CONTROL

* : IF ONLY RACK SENSOR IS DAMAGED, THE ACTUATOR IS CONTROLED BY N FEEDBACK.

| | DUPLICATION ||
COMPONENTS	WIRING ONLY	WIRING & COMPONENTS
BATTERY	○	
ACTUATOR	○	
N SENSOR		○
RACK SENSOR	○	
ACCEL SENSOR	○	

Tab.6 Double-wiring

Tab.5 On-board diagnosis

Engine stalling and over-running are serious conditions, which must be avoided in any failure mode.
The way in which the electronic governor will avoid these conditions, is described in the following.

(1) Engine stalling

Duplication concept is standardized at the following channels.
   Battery
   Actuator
   Speed sensor
   Rack position sensor
   Acceleration pedal sensor
It will be able to decrease the rate of failure caused by miscontact.

(2) Over-running

The system is protected from over-running by sensing the engine speed analogous, i.e. N signal after F-V conversion is compared with the preset value, and if the value exceeds the preset speed, control rack will be moved to zero position using a logical circuit.

PROSPECTS FOR DIESEL ELECTRONIC CONTROL

The system described in this report is the so-called first-generation system of diesel engine electronic control. Although it enables the improvement of fuel ecomony, driveability etc, it does not change the injection performance itself.
If the needs, such as high pressure injection, multi-fueling and varibale injection rate, increase drastically, second-generation system (timed operation with solenoid valve) will be necessary. Up to that time, the use of the first generation system will continue to increase.

ACKNOWLEDGEMENT

We would like to thank Japanese diesel engine makers, which proposed useful needs.

REFERENCES

1. P.A.Hazell and J.O.Flower, "Sampled-data Theory Applied to the Modeling and Control Analysis of Compression Ignition Engines -- Part I" INT.J CONTROL, 1971, VOL.13, No.3

2. J.O.Flower and P.A.Hazell, "Sampled-data Theory Applied to the Modeling and Control Part II" INT.J CONTROL, 1971, VOL.13, No.4

3. S.Suzuki, "The Improvement of Fuel Injection System for Vehicle Diesel Engine Developing High Injection Rate" SAE 790891

4. M.Wakabayashi, S.Sakata and K.Hamanaka, "Isuzu's New 12.0L Micro-Computer Controlled Turbocharged Diesel Engine" SAE 840510

860145

The Second Generation of Electronic Diesel Fuel Injection Systems — Investigation with a Rotary Pump

Keiichi Yamada and Hidekazu Oshizawa
Diesel Kiki Co., Ltd.

ABSTRACT

This paper describes concepts of the next generation of electronic diesel fuel injection (EDFI) systems, and the test results of the prototype, named "Model-1".

Important characteristics of the next generation of EDFI will be ; mechanical simplicity, direct control and pump intelligence.

Direct spill control using a high speed solenoid valve for injection regulation and pump mounted electronic circuits were used in the "Model-1" system.

The test results demonstrate the advantages of this system, and suggest possibilities of new functions such as individual cylinder control, pilot injection and multi fuel usage.

BACKGROUND

Since the early 1980's, DIESEL KIKI Co., Ltd. has been producing micro-computer controlled diesel fuel injection systems - the rotary pump "COVEC"(1)*, and the in-line pump "COPEC"(2)* - which are offering big advantages in fuel economy, emissions and vehicle performance.

In the past few years, requirements on electronic control injection systems have continued to increase, and have become rapidly more complex. Injection systems are being linked with other subsystems such as transmission control, display control, etc., and are also reducing vibration and engine noise without increasing system cost.

---

* Number in parentheses designate references at the end of paper.

SYSTEM CONCEPTS

After reviewing electronic control system requirements for light duty diesel engines, we have concluded that future systems must provide the following advantages over today's systems.

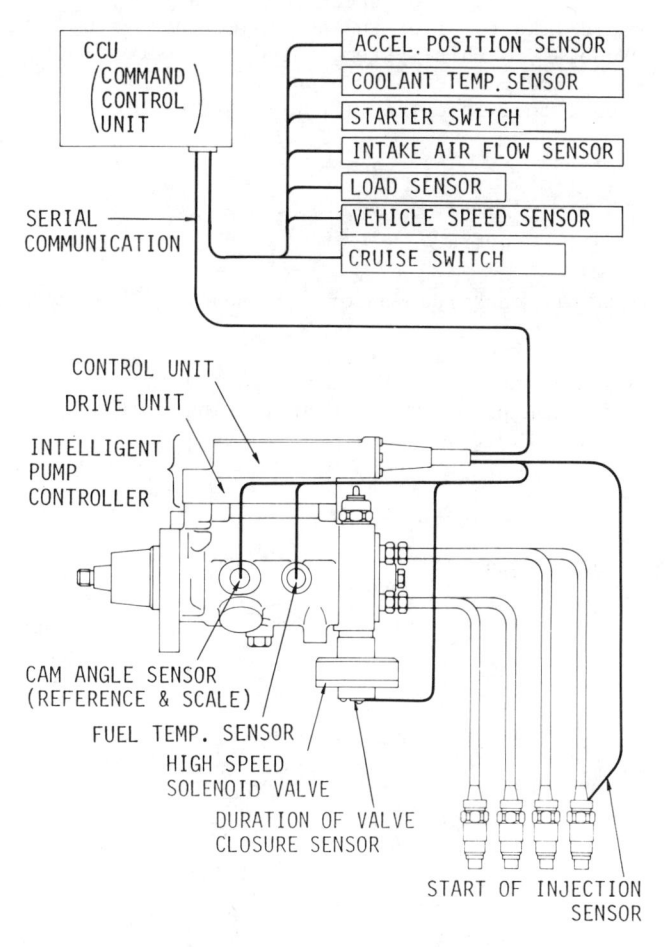

Fig.1 "Model-1" system construction

MECHANICAL SIMPLICITY — The injection control mechanism should be simplified by regulating both timing and fuel quantity using only one actuator.

This will considerably decrease cost and increase reliability.

DIRECT SPILL CONTROL — First generation systems indirectly measure and control fuel injection pressure using analog servo controlled mechanical part positioning techniques.

Fuel injection pressure should be more direcly monitored and adjusted by a microprocessor controlled high speed solenoid valve, using digital control techniques.

This method avoids analog servo inaccuracies, and provides high speed control.

SEPARATE DEDICATED FUEL INJECTION CONTROLLER — If we separate the injection control functions from the main control unit, and integrate them into the pump using another microprocessor, we can provide more powerful specialized control of the pump, and compensate better for the variation of mechanical parts.

This also allows the main control unit to be more flexible and powerful.

As a result of these considerations, we would like to offer the new injection system "Model-1". This system, based on new concepts, will create a new generation of electronic diesel fuel injection (EDFI) systems.

SYSTEM OVERVIEW

The construction of the "Model-1" is shown in Fig.1.

Two functions of the "Model-1" pump are based on the BOSCH VE pump ; generating injection pressure and distributing fuel (Fig.2, Fig.3).

Fig.3 Intelligent pump

A high speed solenoid valve is located in fuel inlet path, and timing and duration of the valve closure regulate injection timing and fuel quantity respectively.

The electronic circuits are divided into the IPC (Intelligent Pump Controller) which is mounted on the pump, and the CCU (Command Control Unit) which is otherwise called the Electronic Control Unit.

The CCU computes fuel quantity and injection timing as functions of engine speed, accelerator pedal position, engine load, and water temperature, etc., then sends commands and data to the IPC.

Receiving commands and data from the CCU, the IPC regulates the injection timing and fuel quantity, and also sends important data, such as pump speed, to the CCU. Moreover, the IPC controls idle speed, and adjusts individual cylinder conditions.

A relationship between the principal events is shown in Fig.4.

Injection timing depends on valve closure (C), and fuel quantity depends on valve closure duration (C-D).

A new type of solenoid "DISOLE" can act very quickly, but it's response time (A-C & B-D) can not be neglected. The response time changes according to manufacturing variations, battery voltage fluctuations, temperature changes, and wear of the mechanical parts.

Therefore, the DVC (Duration of Valve Closure) sensor must detect the timing (C) and (D) of each and every valve opening and closure.

Accurate injection timing (E) is detected using the SOI (Start of Injection) sensor, and drive pulse timing (A) is calculated measuring the duration between (A) and (E).

Pump cam angle determines drive pulse timing and width. Therefore, a cam angle sensor is required to detect cam position accurately. This sensor is also used to measure pump speeds.

The pulse rate of this sensor has been carefully selected so that the controller can accurately compute all pump speeds.

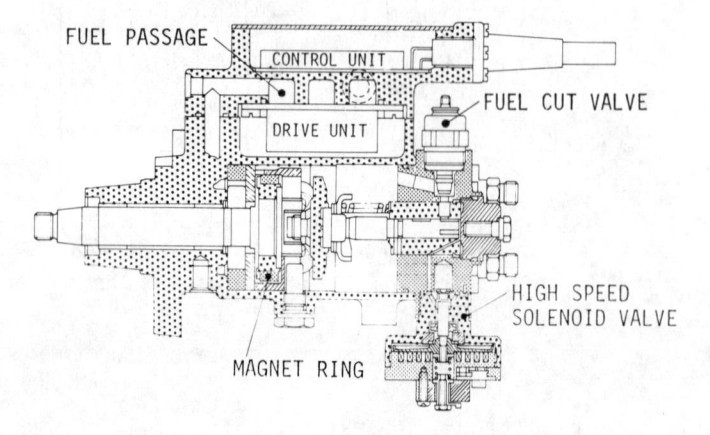

Fig.2 Intelligent pump

To ensure good reliability at high operating speeds, the fast acting "DISOLE" (3)* was developed and adopted for this system (Fig.5, Fig.6).

Fig.4 Relationship between principal events

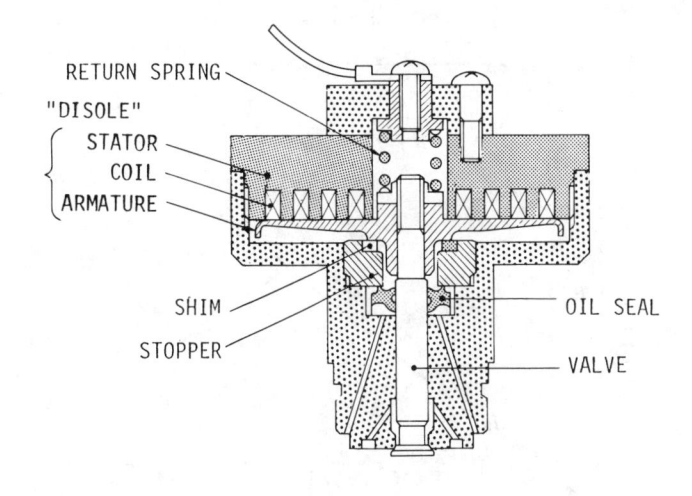

Fig.5 High speed solenoid valve

SYSTEM COMPONENTS

THE HIGH SPEED SOLENOID VALVE — We can estimate the valve motion requirements in the following way:

Supposing a pump has 4 cylinders and the operating speed is 3000 rpm,

$$t_1 + t_2 + t_3 + t_4 \leq 5 \text{ (msec)}$$

where t_1 : response time at closing
t_2 : response time at opening
t_3 : duration of closure at maximum fuel quantity
t_4 : control area of injection timing

Supposing the maximum angle needed to control injection is 35 degrees of cam angle,

$$t_3 + t_4 = 2 \text{ (msec)}$$

Therefore,

$$t_1 + t_2 \leq 3 \text{ (msec)}$$

Fig.6 High speed solenoid valve

Two other important parameters are:
1) maximum fuel flow area at the valve seat,
2) the rate of change of the flow area with valve lift.

A computer simulation was carried out to optimize the valve seat dimensions.

THE CAM PROFILE — Using the conventional BOSCH VE pump cam profile causes undesirable changes in fuel quantity when pump timing is changed.

Therefore, a new cam profile with a constant slope (a constant velocity) was designed (Fig.7).

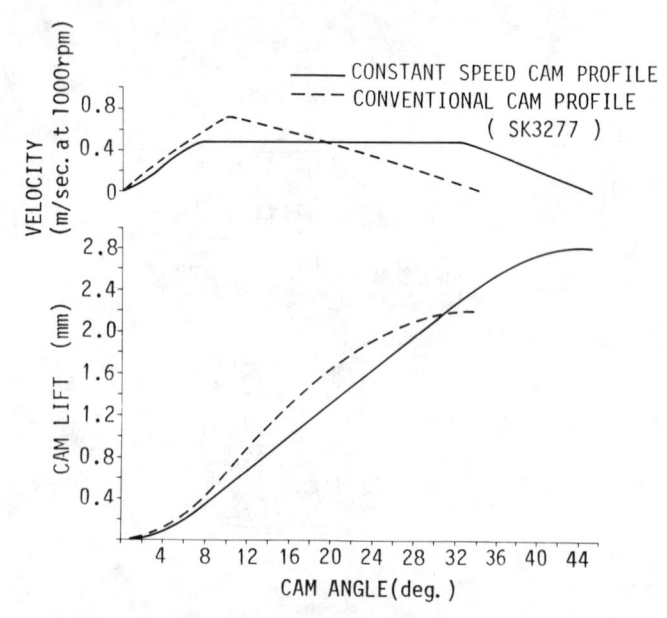

Fig.7 Cam profile

THE SOI (START OF INJECTION) SENSOR — To detect actual injection timing, a contact-point type SOI sensor (4)* was developed.

In this sensor, the sliding surface of the nozzle needle is insulated, and valve seat acts as an electrical contact-point between the nozzle needle and the nozzle body. When the nozzle needle is lifted to start injection, electrical contact is broken and highly accurate start of injection timing can be detected(Fig.8). The highly durable ZrO_2 insulation coating is manufactured using an ion plating process.

THE DVC (DURATION OF VALVE CLOSURE) SENSOR — To detect actual timing of valve closing and opening, a similar contact-point type DVC sensor, using the same ZrO_2 insulation coating, was designed (Fig.9).

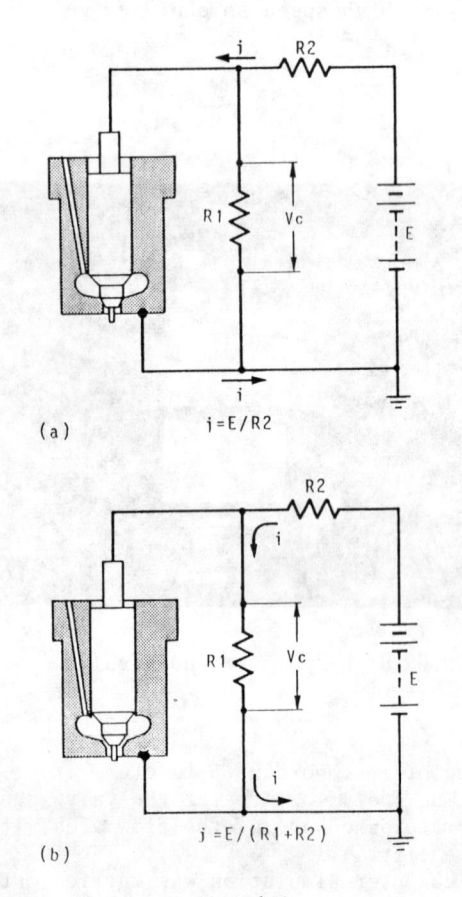

Fig.8 SOI sensor model

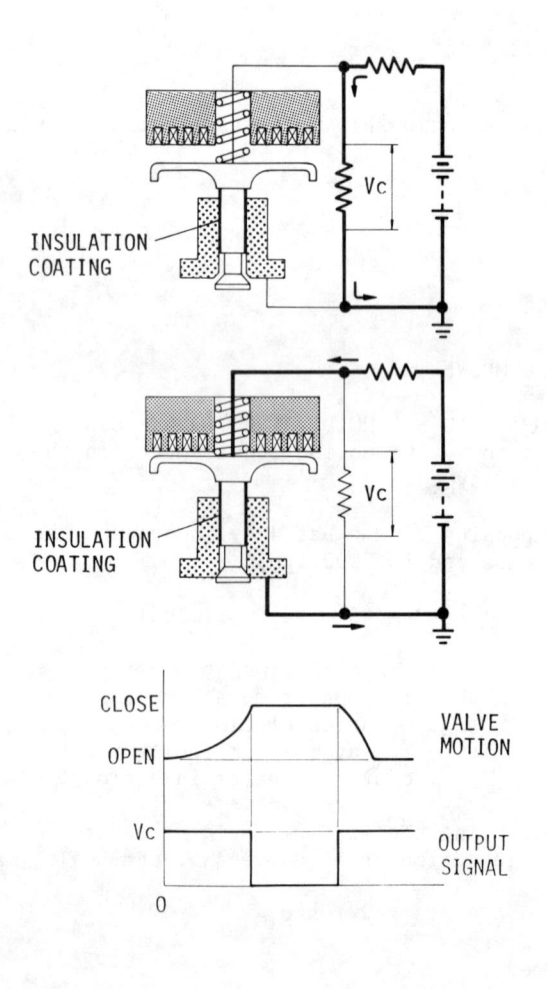

Fig.9 DVC sensor model

THE CAM ANGLE SENSOR — To accurately detect cam angle even at very low starting speeds, a Hall element and magnet ring design was adopted.

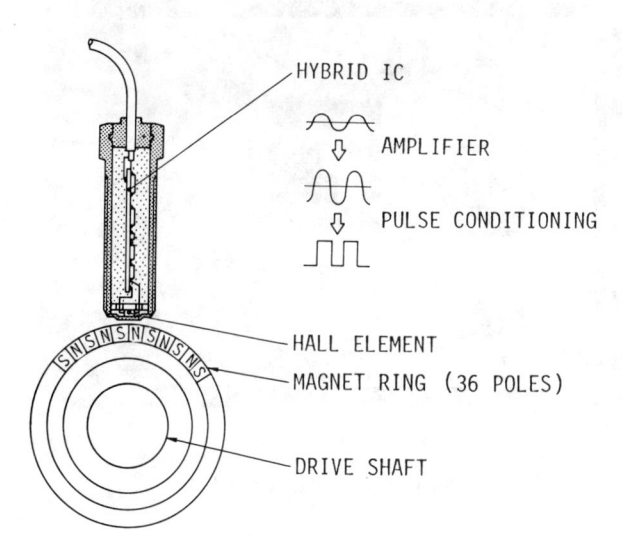

Fig.10 Cam angle sensor model

Fig.11 Cam angle sensor

This sensor allows easier initial adjustment of the relationship between the pump camshaft and the crank shaft when installing it on the engine.

A hybrid IC is used to integrate amplifier and signal conditioning circuits in the sensor because the Hall output signal amplitude is very low (Fig.10, Fig.11).

THE IPC —

Control Functions —
* Driving the "DISOLE" to control injection according to physical data received from the CCU.
* Sensing pump speed, and sending it to the CCU.
* Compensating the fuel quantity of individual cylinders.
* Storing calibration data which allows injection pump usage on many engine types as well as allowing compensation for specific pump variations.
* Updating the calibration data, when the pump is being installed on the engine or being serviced in the field.

Configurations — The IPC has two modules. One is the control circuit which computes complex functions, the other is the drive unit which regulates high current in the "DISOLE".

Both electronic circuits are integrated on small ceramic base boards and mounted on the pump.

To ensure high temperature reliability, the circuits are cooled using fuel (Fig.12).

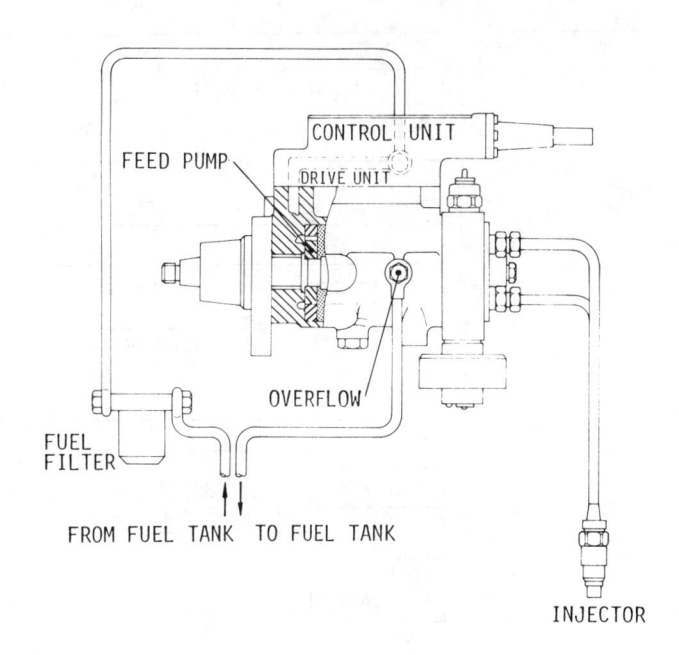

Fig. 12 IPC cooling using fuel

The Control Circuit -- A schematic of the circuit is shown in Fig.13.

The circuit is integrated on a 63cm² size ceramic base board, and consists of four C-MOS LSI's which are packaged in PLCC's (Fig.14).

These LSI's are the central processing unit (CPU), the erasable programmable read only memory (EPROM), the analog to digital converter (ADC), and the signal processing unit (SPU).

The SPU is a gate-array, and performs many important functions ;

1) Handling sensor signal such as the DVC, SOI, the cam angle signal, etc..
2) Measuring the pump speed and the response time of the valve and the SOI signal.
3) Detecting the cylinder firing order.
4) Outputting the solenoid drive pulse at the requested timing with accurate pulse width.
5) Decoding chip select signals for peripheral LSI's.
6) Latching address data from the CPU multiplex bus.

Calibration data is stored in the EPROM. The EPROM will be upgraded to an electrically erasable PROM (EEPROM) in the near future.

Fig.14 Control unit hybrid IC

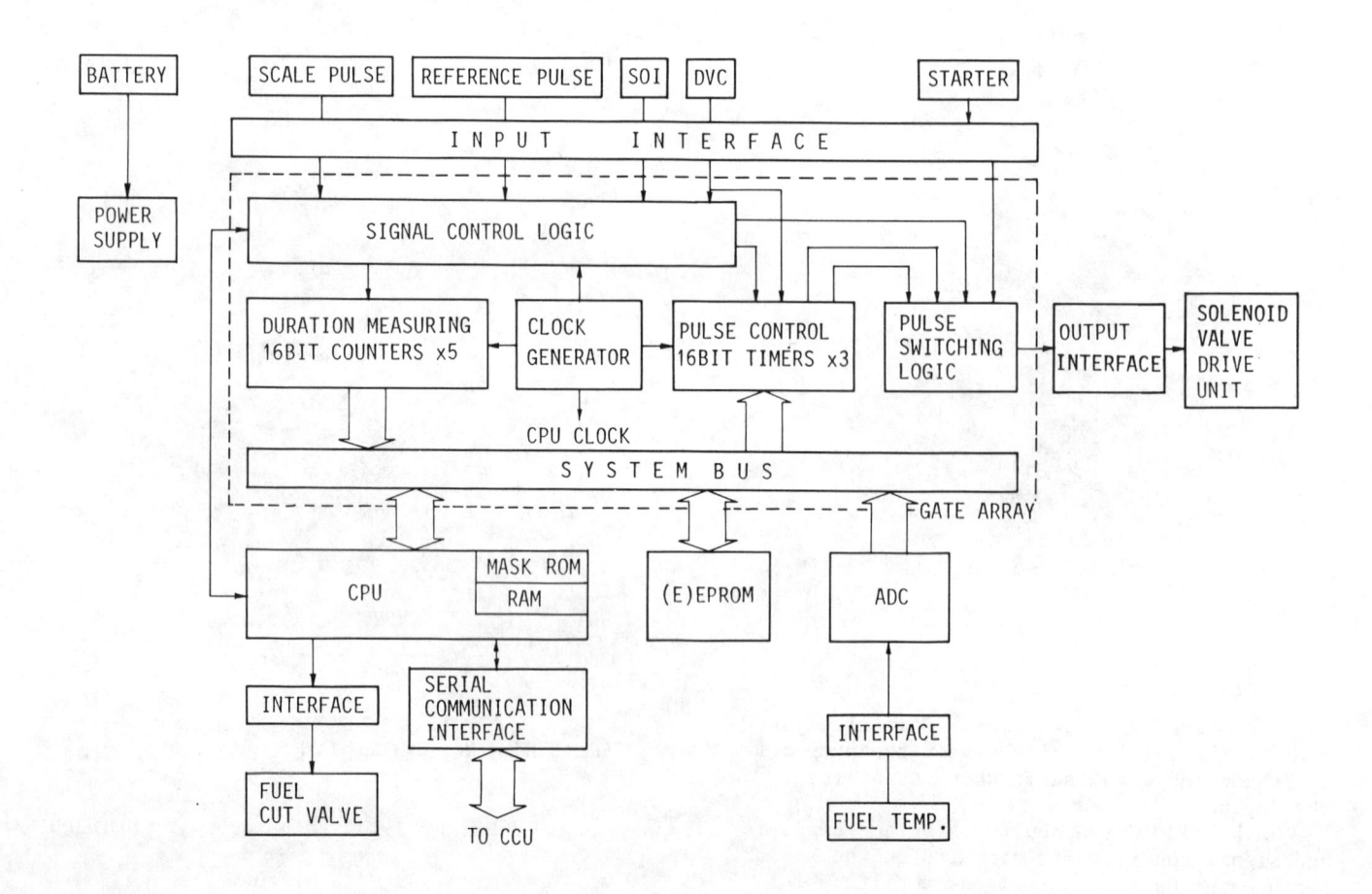

Fig.13 Control unit schematic

THE DRIVE UNIT — This unit regulates the current in the "DISOLE" to maximize actuator speed while minimizing power dissipation (Fig.15).

THE CCU — A basic schematic of the CCU is shown in Fig.17.

In the "Model-1", the CCU was designed for fundamental functions only. Many design approaches are possible to meet various engine and vehicle requirements.

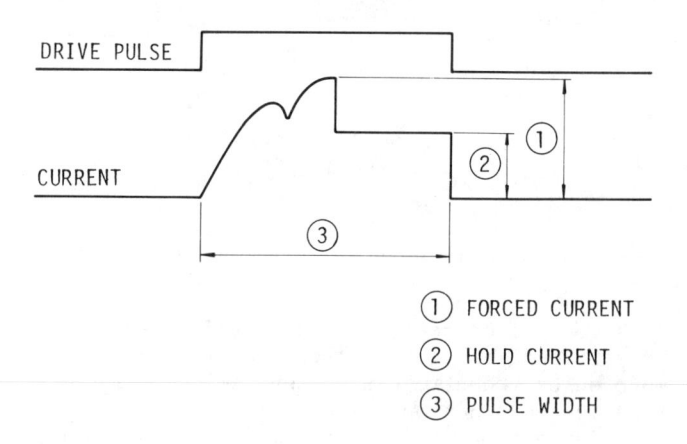

① FORCED CURRENT
② HOLD CURRENT
③ PULSE WIDTH

Fig.15 Current regulating model

It consists of a current regulating circuit integrated on a 10cm^2 size ceramic base board, and a high power transistor bare chip (Fig.16).

Fig.16 Drive unit hybrid IC

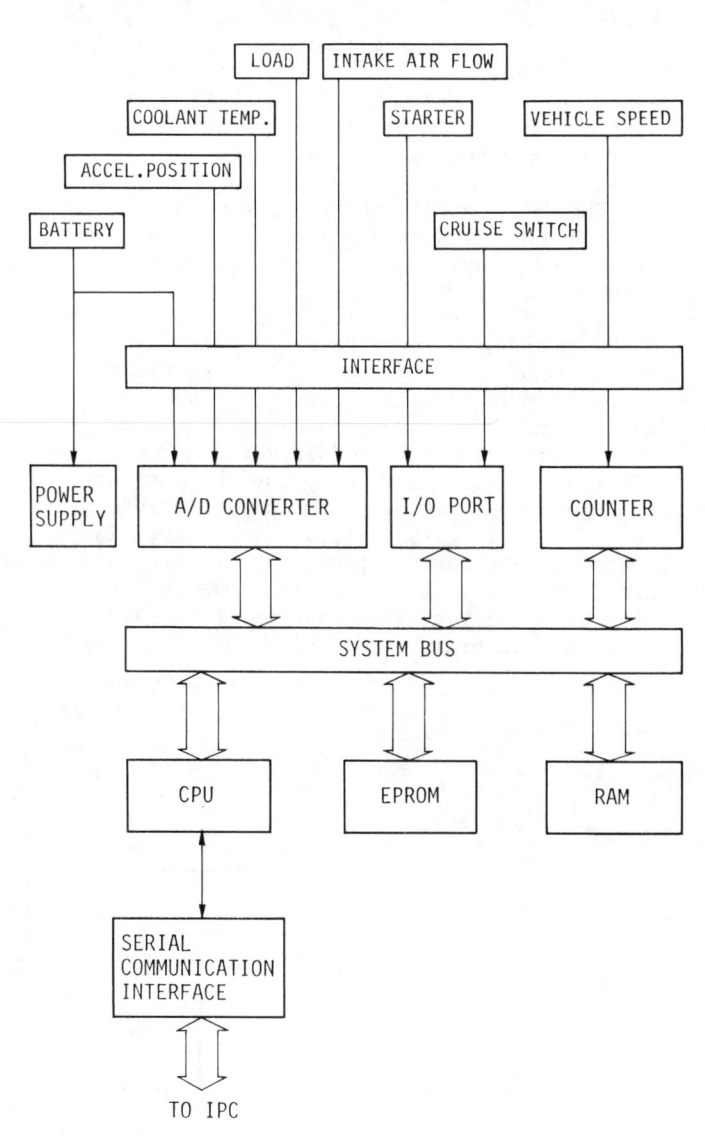

Fig.17 CCU schematic

THE SERIAL LINK — The serial link between the CCU and the IPC has a very high speed transfer rate, and is isolated by photocouplers.

Data communication is performed in our original protocol using packets of command and data.

CONTROL DESCRIPTION

INJECTION TIMING CONTROL -- The optimized injection timing data is stored in the CCU ROM as a look-up table. The CCU determines the optimum injection timing from the table as a function of engine speed and load conditions, and sends the data to the IPC as a reference crank timing value.

Drive pulse is computed as shown in Fig.18.

The IPC converts the reference timing data from crank angle to cam angle using calibration data. This calibration data is the specific relationship between crank angle and pump cam position when the pump is installed on an engine.

The IPC controls the valve closing timing to match the SOI signal and the converted reference timing data. The control algorithm is as follows.

The IPC controls drive pulse output timing to the solenoid valve. However, there are several time delays between the pulse timing and the actual injection timing, such as the pressure build up time in a plunger chamber, and pressure propagation delays through the injection pipes. Therefore, the IPC must determine the output timing of the pulse including these delays (Fig.19). Total delay time (Tsdi) is affected by engine speed, temperature, and injection pipe length. The IPC measures this delay time every cycle, converts it to cam angle, and determines the pulse output timing.

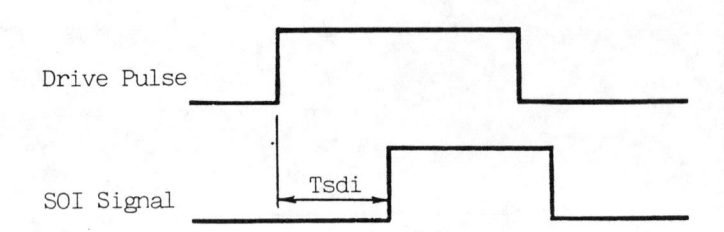

Fig.19 Relationship between drive pulse and SOI signal

This control concept, when compared to the proportional-and-integral control method, can more quickly control injection timing during transient engine conditions.

For optimum injection timing control during this process, the pump cam angle must be detected accurately. The cam angle sensor generates signals exactly at every 10 cam degrees, and the period between each signal is measured in time dimension to compute fine cam angles of less than 10 degrees.

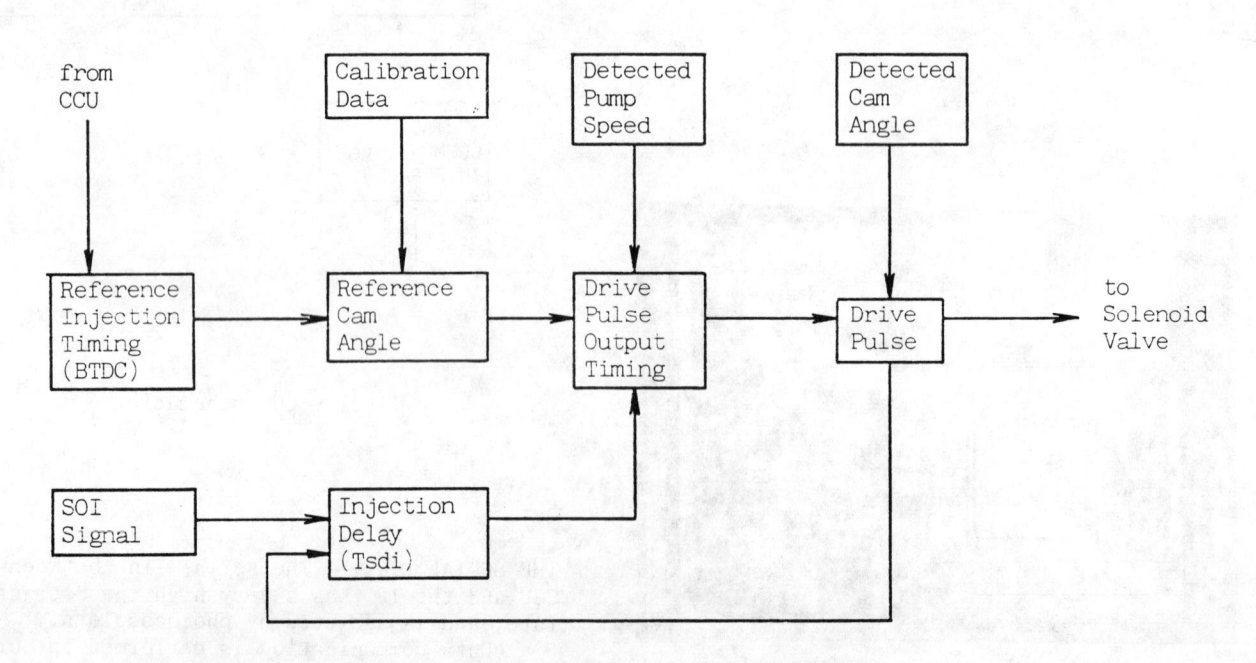

Fig.18 Injection timing control flow diagram

FUEL QUANTITY CONTROL — The CCU computes an optimum fuel quantity as a function of engine speed and accelerator pedal position, and sends it to the IPC as a reference fuel quantity. This method of determination is the same as that used for injection timing control.

Drive pulse width is computed as shown in Fig.20.

The IPC modifies the received data, and adjusts individual cylinder fuel quantities to compensate for fuel temperature changes and individual cylinder and pump conditions.

The compensated fuel quantity data is converted to the reference valve closing duration cam angle (Avref) using a calibration data table. This calibration data table provides the specific relationship between required fuel quantity and valve closing duration angle as a function of pump speeds.

The valve closing duration is reconverted from cam angle to duration timing (Tvref) using instantaneous pump speed data to determine the drive pulse width.

The drive pulse width (Td) must include valve closing and opening response times (Tsdv & Tedv). To detect valve closing and opening, the DVC sensor is used (Fig.21).

The IPC computes the drive pulse width as follows.

$$Td = Tsdv + Tvref - Tedv$$

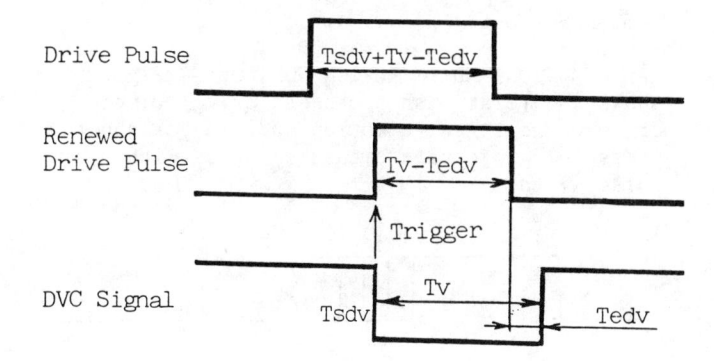

Fig.21 Relationship between drive pulse and DVC signal

After outputting this pulse, the IPC renews the drive pulse width.

$$Td = Tvref - Tedv$$

The renewed drive pulse width is triggered by the DVC signal's leading edge.

Using this renewed drive pulse offers two advantages :

1) Valve closing delay effects become negligible.
2) The most recent pump speed data becomes available for converting cam angle to valve closing duration timing.

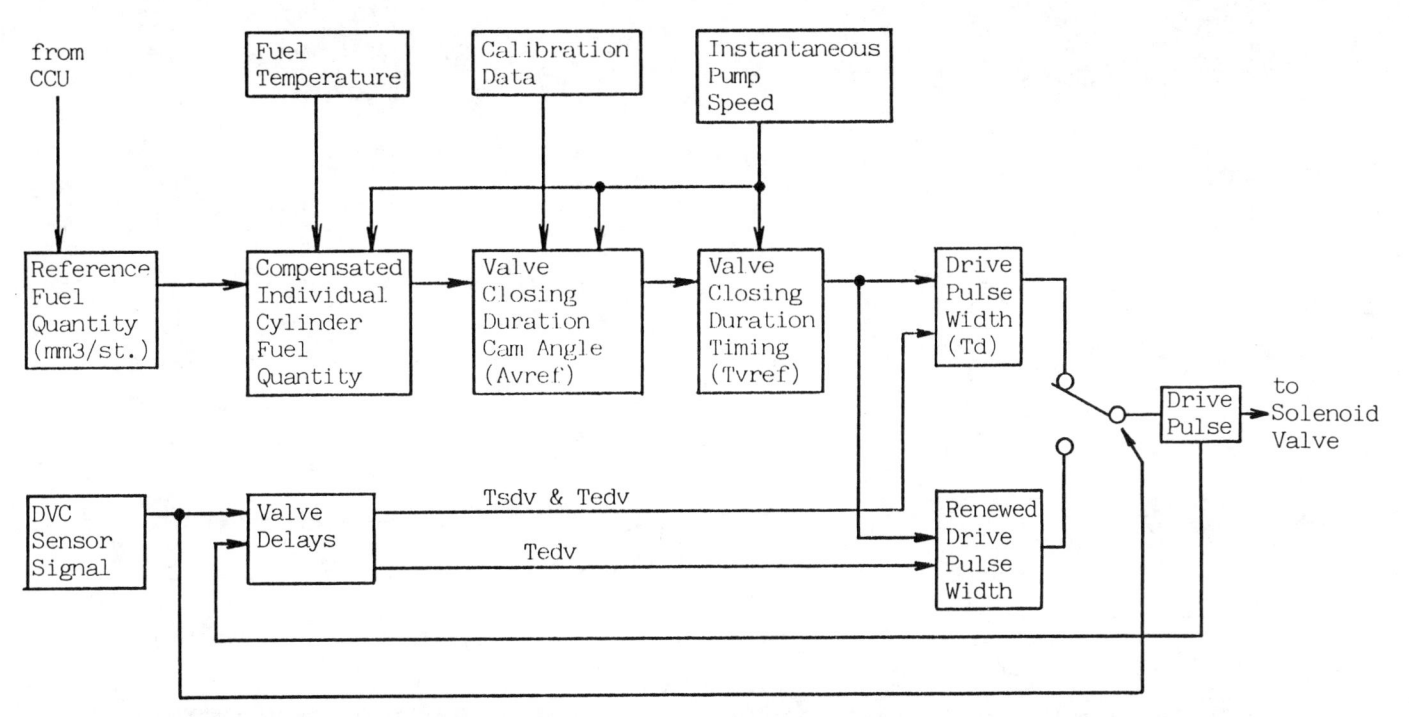

Fig.20 Fuel quantity control flow diagram

TEST RESULTS

VALVE MOTION AND INJECTION CONTROL — Fig.22 shows the relationship between drive pulse, drive current, valve motion and plunger chamber pressure. Injection pressure varies with drive pulse width.

Valve motion is considerably influenced not only by solenoid force but also by the spring constant and the force setting of the return spring. Therefore, suitable values were chosen after considering injection controllability and stability.

Fig. 23 shows that fuel quantity is injected in proportion to the duration of valve closure.

INJECTION TIMING CONTROL — Fig. 24 shows fuel quantity v.s. valve closure timing under constant duration of valve closure conditions.

When using the conventional VE pump cam profile, fuel quantity is adversely affected as injection timing changes (upper diagram in Fig. 24).

This problem was avoided by using the constant velocity cam profile which allows fuel quantity and injection timing to be independent (lower diagram in Fig. 24).

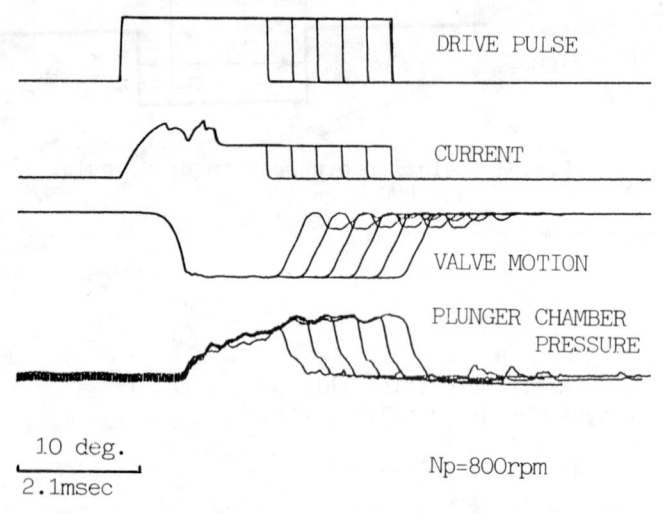

Fig.22 Relationship between electrical events and injection

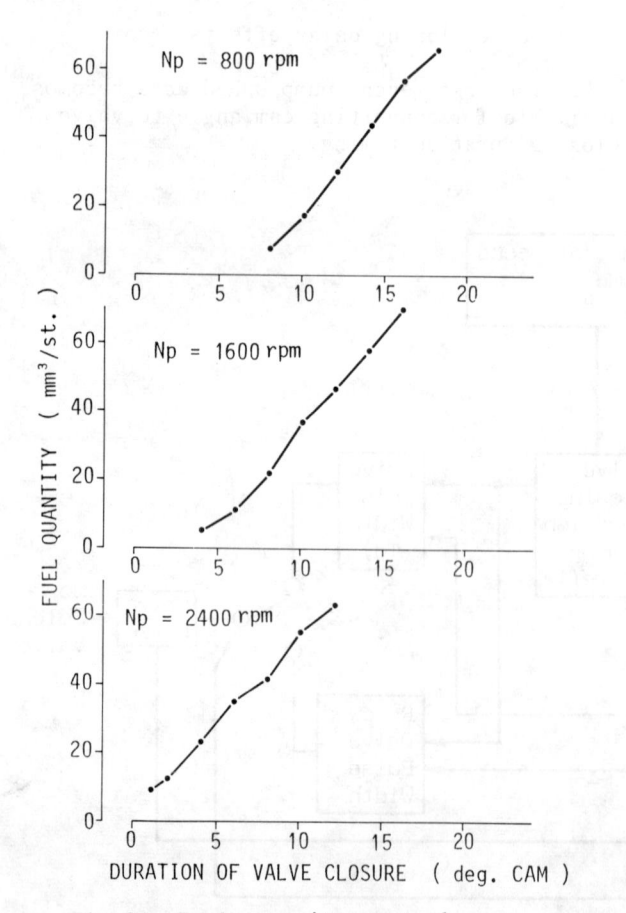

Fig.23 Fuel quantity control

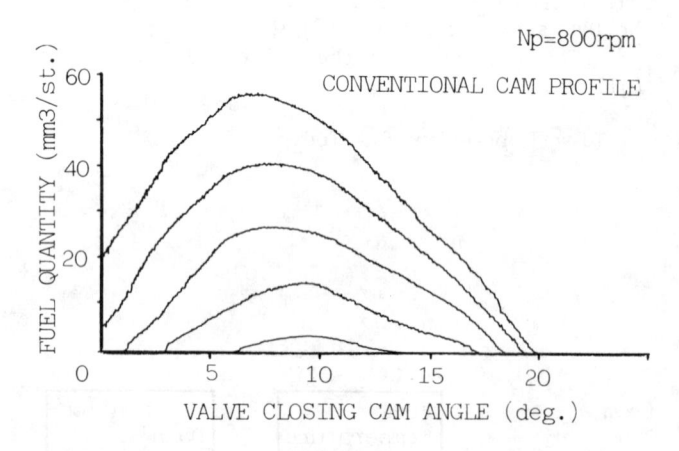

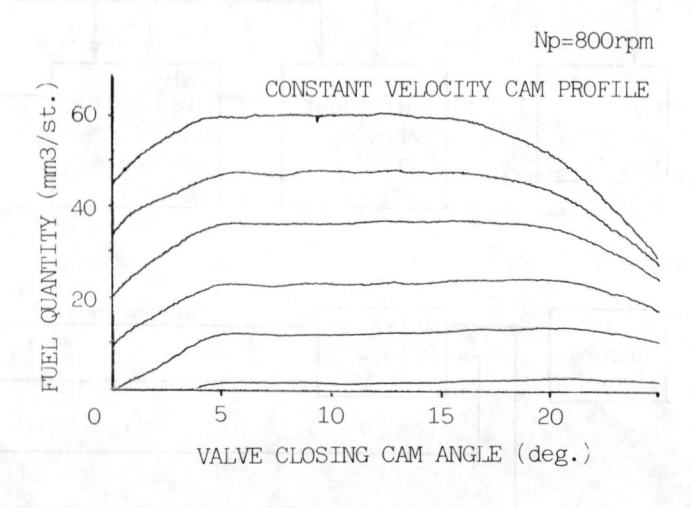

Fig.24 Injection timing control

IPC COOLING USING FUEL — Fig. 25 shows that electronic circuit temperature rise is kept within the desired reliability range.

FUEL TEMPERATURE AT INLET ; 60°C
DRIVE PULSE DUTY CYCLE ; 30 % AT 160 Hz
(4cyl. 60mm³/st. 2500 rpm)

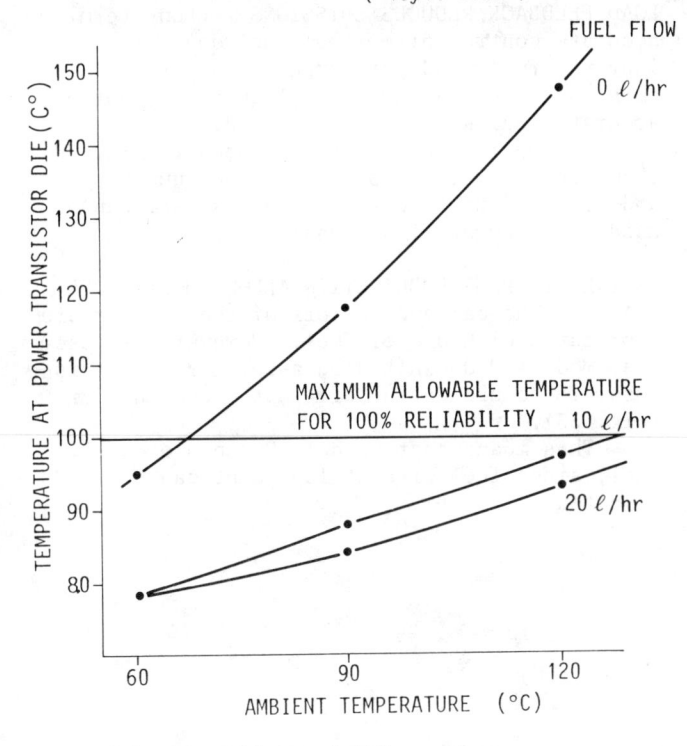

Fig.25 Effect of IPC cooling

DVC SENSOR ADVANTAGES — In Fig. 26, the solid line shows the result of fuel quantity control using the DVC sensor signal. The dotted line shows the result of fuel quantity control using pre-determined values for valve closing and opening delays.

DVC sensor use provides more accurate fuel quantity control.

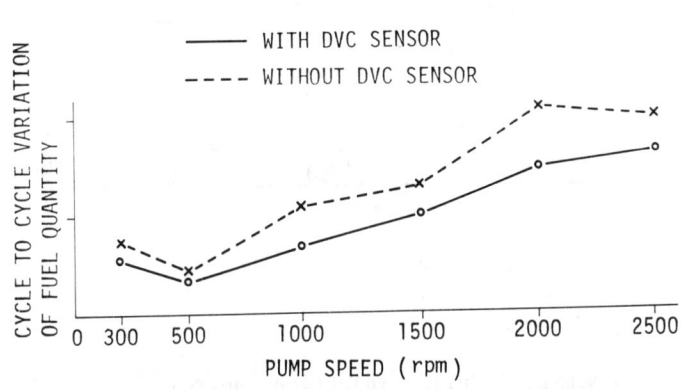

Fig.26 Fuel quantity control using DVC sensor

INDIVIDUAL CYLINDER CONTROL — Engine testing conditions are shown in Fig. 27.

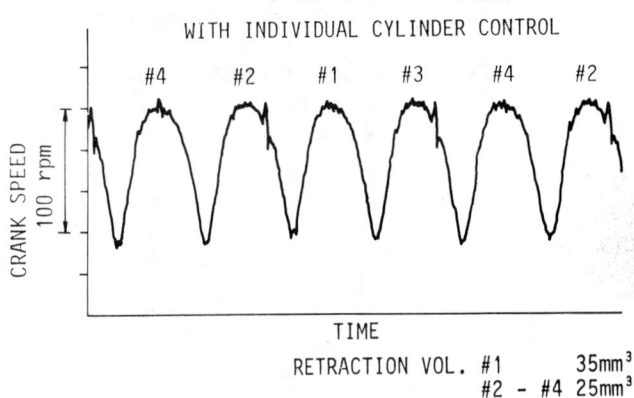

Fig. 28 shows individual cylinder crank speed variation with and without individual cylinder control. Fig. 29 shows the crank speed power spectrum with and without individual cylinder control, and Fig. 30 shows the similar vibration acceleration power spectrum.

Individual cylinder control reduces uncomfortable low frequency vehicle vibrations.

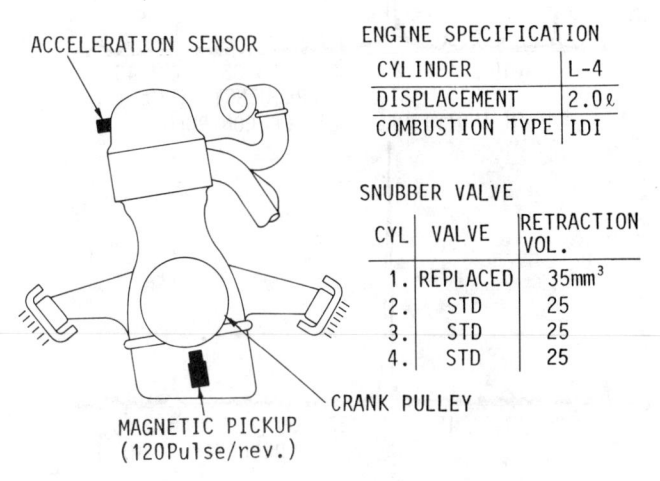

ENGINE SPECIFICATION

CYLINDER	L-4
DISPLACEMENT	2.0ℓ
COMBUSTION TYPE	IDI

SNUBBER VALVE

CYL	VALVE	RETRACTION VOL.
1.	REPLACED	35mm³
2.	STD	25
3.	STD	25
4.	STD	25

Fig.27 Engine test conditions

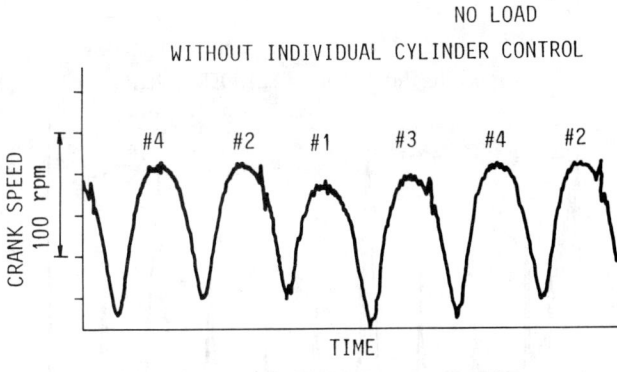

Fig.28 Individual cylinder crank speed variation

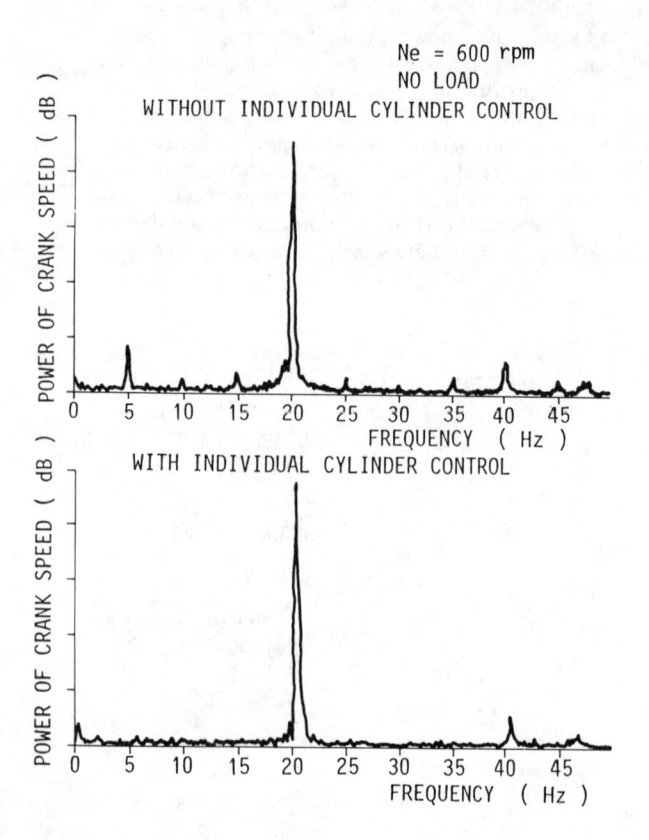

Fig.29 Power spectrum of crank speed

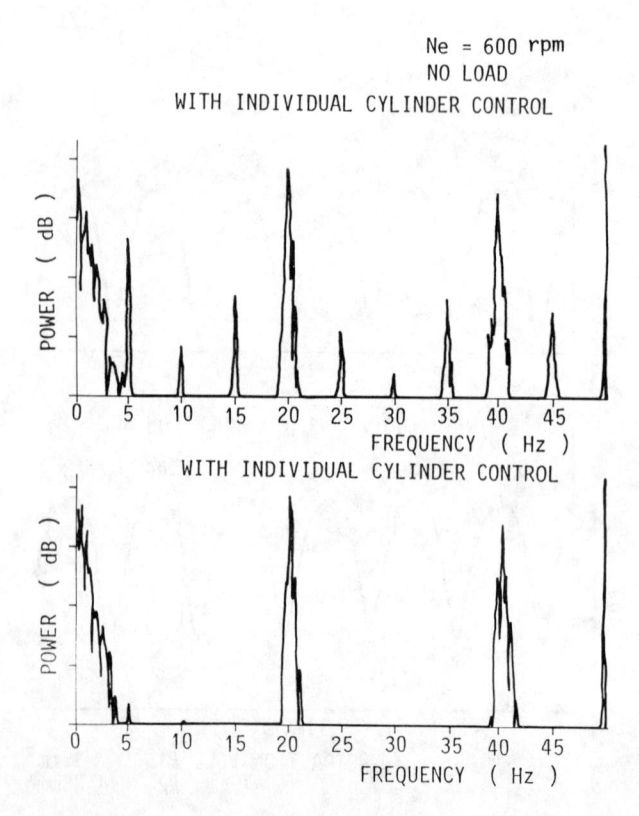

Fig.30 Power spectrum of vibration at cylinder head

ADDITIONAL ADVANTAGES

PILOT INJECTION REDUCES AUDIBLE NOISE — If pilot injection can be utilized, premixed combustion reduction will decrease audible combustion noise.

Fig. 31 shows pilot injection with the "Model-1".

LOAD FEEDBACK REDUCES EMISSIONS — Long term accurate control of maximum fuel quantity improves full load performance and particulate emissions. For accurate fuel quantity control, accurate load sensing is required.

The performance of the O_2 sensor (Fig. 32) shows the relationship between fuel quantity and sensor output voltage. This sensor can be used for accurate load sensing.

ENGINE OIL PUMP LUBRICATION ALLOWS MULTI-FUEL USAGE — The cam and rollers of the VE pump are lubricated with diesel fuel. However, because the "Model-1" doesn't need a spill ring, it is possible to use engine oil in the cam chamber (Fig. 33).

This means that "Model-1" can use multi-fuel, since fuel without lublicant can be injected.

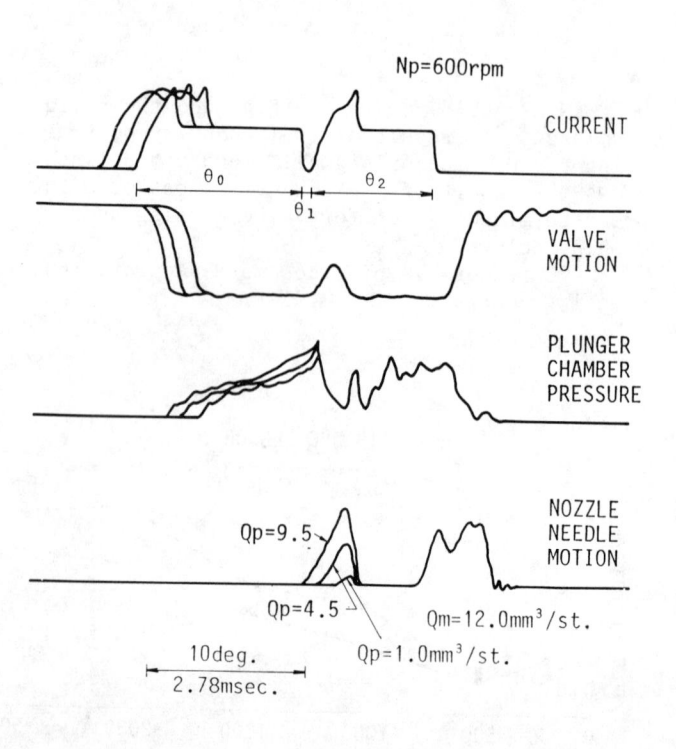

Fig.31 Pilot injection control

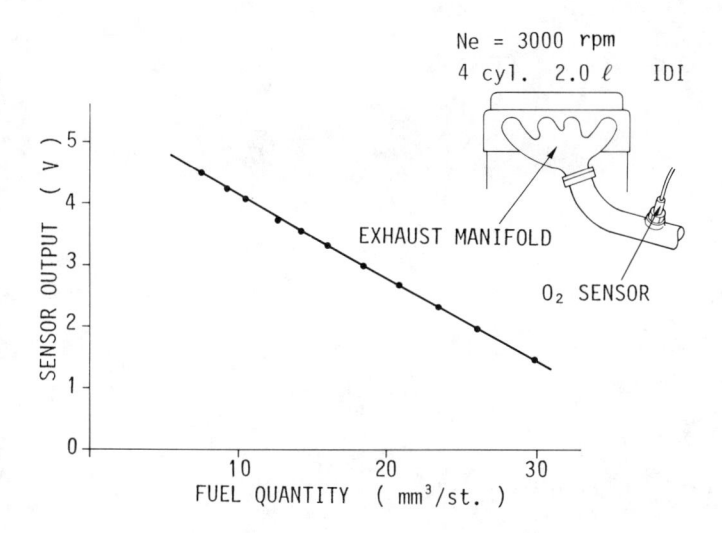

Fig.32 O₂ Sensor output

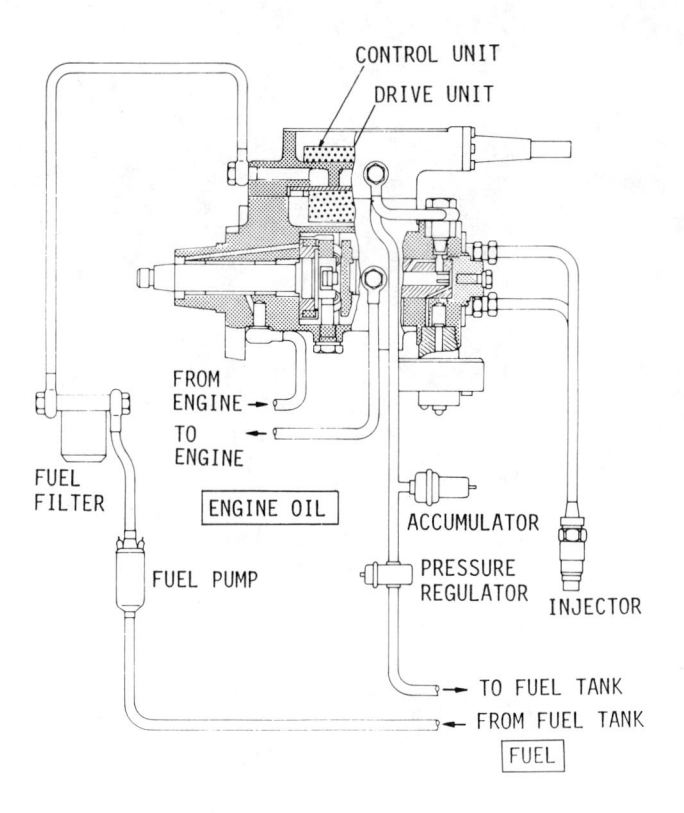

Fig. 33 Engine oil pump lublication

SUMMARY

[1] Direct spill control using a high speed solenoid valve for a rotary pump has been successfully demonstrated. This allows mechanical simplification and direct digital control.

[2] In order to realize direct spill control, a high speed solenoid "DISOLE", a cam angle sensor, and a DVC sensor have been developed.

[3] A separate dedicated fuel injection controller was integrated into the pump design. It was confirmed that electronic circuits can be reliably cooled using fuel.

[4] This system offers flexibility ; as a result of using divided circuitry, one pump model can easily be applied to a wide range of engines and vehicles.

[5] The "Model-1" will lead to a new generation of EDFI systems offering these advantages ;
   1) Reduction of vehicle vibration through individual cylinder control.
   2) Reduction of engine noise by pilot injection.
   3) Improvement of emissions and full load performance.
   4) Usage of multi fuels.

REFERENCES

(1) R. Kihara, Y.Mikami and H.Nakano
"The Performance Advantages of Electronic Control Diesel Engine for Passenger Cars" SAE830528
(2) M. Wakabayashi, S. Sakata and K. Hamanaka
"Isuzu's New 12.0L Micro-Computer Controled Turbocharged Diesel Engine" SAE840510
(3) T. Kushida
"High Speed, Powerful and Simple Solenoid Actuator "Disole" and its Dynamic Analysis Results" SAE850373
(4) M. Kasaya, and T. Abe
"A Contact-Point Type Start of Injection Sensor for Diesel Engines" SAE851585

860146

The Electronic Governing of Diesel Engines for the Agricultural Industry

Peter Howes, David Law, and Dalip Dissanayake
Lucas CAV Ltd.

ABSTRACT

This paper explores the governor characteristics which can be provided for engines in tractors and other off highway applications. These do not have to meet stringent emission regulations, and therefore do not need precise control of timing and maximum fuel.

The paper describes a simple electronic governor concept which has two regions, one for driving on the road and the other giving close, all speed regulation for off highway and Power Take Off operations.

Such an approach offers better governing, driveability and communication with other vehicle systems. In the field this would give a more uniform rate of working and spreading of fertilizers etc, less tendency to stall and result in a more economic operation.

This system replaces the standard mechanical governor, but otherwise the pump remains unaltered, so the engine manufacturer can fit either the mechanical or electronic governor as required without additional engine application work or change to its power rating.

This technique could be used on industrial engines and generator sets.

ELECTRONIC CONTROL of diesel engined Road Vehicles is now being introduced into production. Various systems have been described, all of which control maximum fuelling, injection timing, torque curve shape and governing (refs: 1, 2, 3, 4, 5). The main stated advantages of these systems are:

1. Improved fuel economy.
2. Lower exhaust emissions.
3. Improved driveability.
4. Improved diagnostics and fringe benefits linking in to a total vehicle electronic system.

On Small Off Highway vehicles, for example agricultural tractors, such total electronic control is not necessary.

It was considered, however, that such tractors, of less than 100 HP, would still benefit from the improved driveability and implement control obtainable with electronic governing.

Frequently, for some operations, close governing has been called for, but with a mechanical governor, it is difficult to provide close governing, that is 5% or less, at all speeds. Furthermore, if it could be done, the driveability on the road would be poor.

If electronic control of governing can be provided, it should be possible to have both a close governor and a shallow governor, the choice selected on a control panel.

Electronic governors for generator sets are on the market today, which are bolt on devices operating through the normal pump throttle lever and mechanical governor. The use of electronic governing applied directly to the control means (ie the Metering Valve), will provide the required control in the most economic way.

The aim of the feasibility study discussed later in this paper therefore was to provide an electronic governing system which would replace the mechanical governor, thus providing all the governing advantages of a full electronic system for much reduced complexity. The application range of the pump would therefore be extended.

OBJECTIVES

A specification for the simplest governor for agricultural and industrial engines could be:

FOR TRACTORS - To provide a governor having two regimes, one for driving on the road and the other for Power Take Off on Off Highway use selected by a switch on the control panel.

Road Speed Governing (RS) - To give good "on the road driveability", and yet have better low speed droop characteristics than a mechanical governor.

<u>Off Highway or Power Take Off Governing (PTO)</u> - To be an all speed governor, with constant droop at all speeds of not more than 5% between 1000 rpm and maximum speed.

FOR GENERATOR APPLICATION -

<u>ISO Class A</u> - To provide a governor to enable engines to meet the ISO 3046/IV Class A, governing specification at 1500 and 1800 rpm..

<u>Isochronous Governing</u> - To provide a governor capable of controlling the engine at constant speed, zero droop.

ADAPTABILITY - All of the above to be achieved by modification to an existing mechanically governed pump, with the ECU being the only different unit between tractor and generator sets. The existing hydraulic characteristics of the pump remaining unaltered.

SERVICE CONVERSION KIT OPTION - To provide a system which could easily be fitted in production, or as a service conversion kit, without having to re-match the hydraulics of the pump to the engine.

RELIABILITY - To make the system as simple as possible with minimum additional electronics to ensure maximum reliability.

FAIL SAFE - In the event of any component failing, this must result in the system shutting down.

DESCRIPTION OF SYSTEM

The DPA rotary pump is the most commonly used injection pump on low horse power tractors and was therefore chosen as the fuel injection pump for the evaluation of such a governor system.

The system block diagram is shown in Fig 1

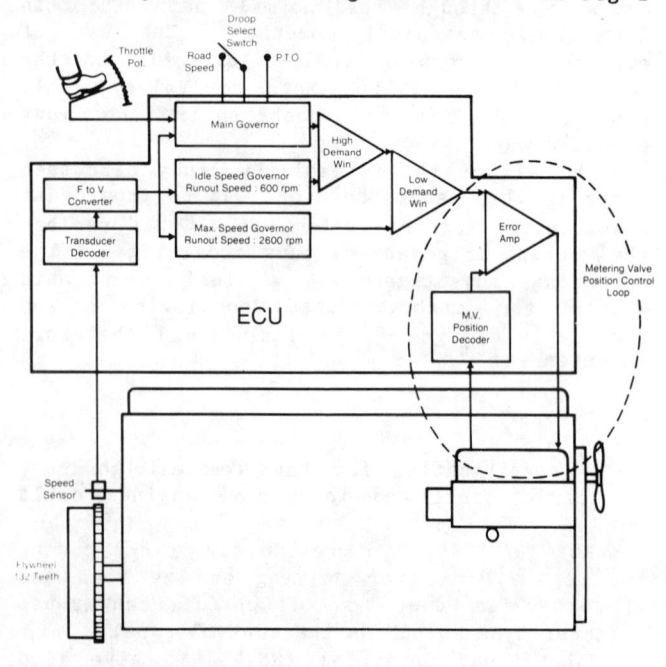

Fig. 1 - System Block Diagram

and a photograph of the total system is also included.

Instead of the mechanical governor, the metering valve was operated by a linear solenoid, connected to a position transducer, all under the governor cover, and an engine speed transducer was fitted on the flywheel.

A speed decode circuit was then arranged to convert the frequency from the engine speed transducer to a voltage signal and supply this to the governor parts of the ECU. Thus when, for example, the Road Speed Governor was in operation, this signal was compared to the demand from the throttle potentiometer and the difference between them was then modified by proportional and integral control terms, amplified and fed into the metering valve control loop, thus repositioning the metering valve.

In this system therefore, if a speed greater than the maximum engine speed was demanded by the pedal potentiometer, a separate maximum speed governor was provided to take over by means of the level of a "which wins" gate. Similarly, a separate idle governor was provided to take over if a lower than idle speed was demanded by the pedal transducer, again via a "which wins" gate.

A further intermediate governor was provided to give close all speed governing for Power Take Off or other Off Highway operations which required a tight droop governor. This was designed to operate in a similar fashion to the Road Speed Governor, but adjusted for tighter control. Both of these intermediate governors were optimised for good driveability and transient performance.

The steeper droop governor was arranged to be selected by moving the control panel switch to PTO (Power Take Off). This was not operational below 1000 rpm, and if the engine speed fell below this level when this mode had been selected, the governing reverted back to road speed governing as the steeper droops were not appropriate at low speeds.

A safety overspeed trip was included such that should the engine speed exceed a set figure the "stop" solenoid was de-energised, thus stopping the engine.

Similarly, the stop solenoid is de-energised in the absence of a speed signal.

To test out the feasibility of this governor a prototype was built.

FIRST RESULTS

These are shown in Figure 2, where the slopes of both the Road Speed Governor and the Power Take Off Governor are shown. (In this terminology, the Power Take Off (PTO) is used to illustrate that governor which is intended for Off Highway use and not just for power take off use).

In this first attempt the Road Speed Governor was set to have around 7% at maximum speed, with a similar slope at the lower speeds. The result was that less than 4% was

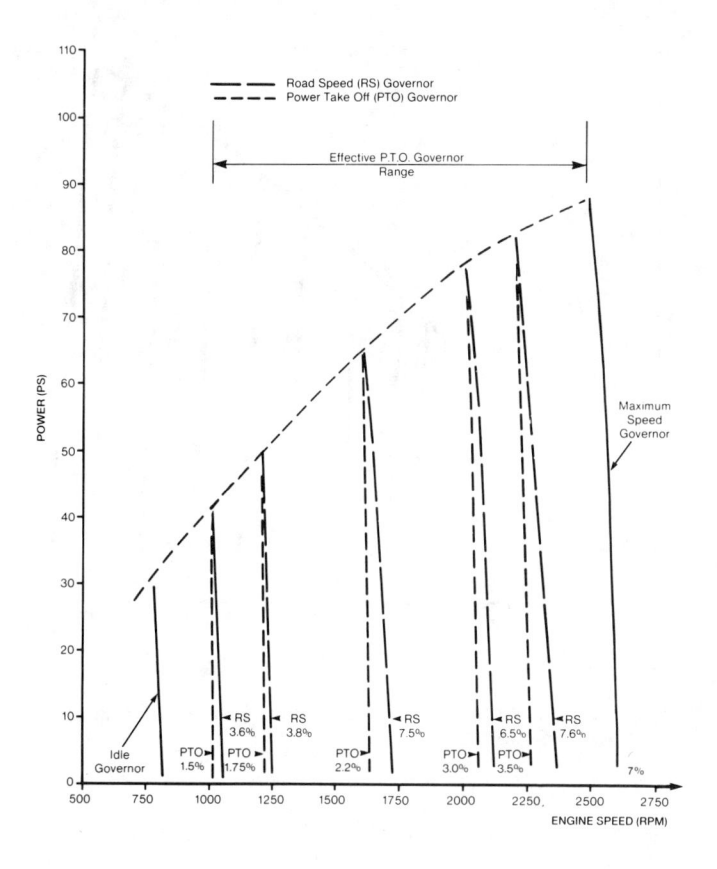

Fig. 2 - EGDPA Engine Governor Pull Offs (First Results)

obtained at low speeds, and hunting below 800 rpm at certain loads became a problem.

The Power Take Off Governing Tests indicated that less than 4% could be maintained between 1000 and 2200 rpm. Indeed, at generator speeds, the governor could be adjusted to give extremely close governing.

This system, whilst showing early promise, suffered from the following problems:
1. Low speed hunting.
2. Drift of speed with temperature.
3. Unsatisfactory transient response ie. too much undershoot on deceleration to idle recovery and too much overshoot on sudden removal of 100% load.

However, in spite of these shortcomings, the system was fitted on to a tractor, to obtain an initial response from an operator. The tractor was put to work in a field and driven by three different people in the power take off mode, with a very satisfactory outcome. All commented on how responsive the tractor was, their main points being:
1. Very good response to throttle changes.
2. Good antistall characteristics resulting in the ability of the tractor to pull itself out of a sticky situation which would have stalled out a tractor with a mechanically governed engine.
3. The ability of the tractor to move over the ground at a more uniform speed, uphill and downhill.

Thus encouraged, further development work was authorised to overcome the known problems.

REVISED SYSTEM TO OVERCOME PROBLEMS

TEMPERATURE DRIFT - This was traced to the temperature sensitivity of the Metering Valve position transducer, so the circuit was modified to compensate for temperature.

HUNTING AT LOW SPEED - This was the most serious problem. The problem really stemmed from the fundamental metering valve characteristics, shown in figure 3, combined with trying to obtain the steeper droops over a wide speed range.

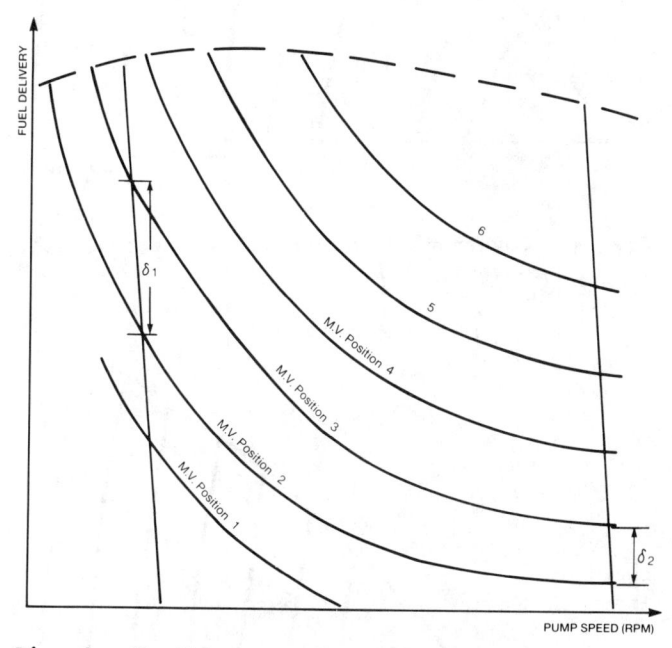

Fig. 3 - Fuelling versus Pump Speed for various Metering Valve positions

A given change of movement at high speed results in a given change of fuel level, but the same movement at low speed results in a much greater change of fuel level. Hence the system is much more sensitive at low speed to metering valve movement.

It was therefore necessary to modify the control circuit and vary gain over the speed range, and at the same time slacken off the low speed droop.

On the Road Speed Governor this was considered desirable anyway for good on road driveability.

The results are shown in Figures 4, 5 & 6. Although the low speed Road Speed governing slope has been increased, it is still better than that with the mechanical governor.

The Power Take Off Governor still gave less than 5% at any speed between 1000 and 2400 rpm.

Additionally, good stability was obtained, as is shown in the figures referred to above, and during the transient response work now described.

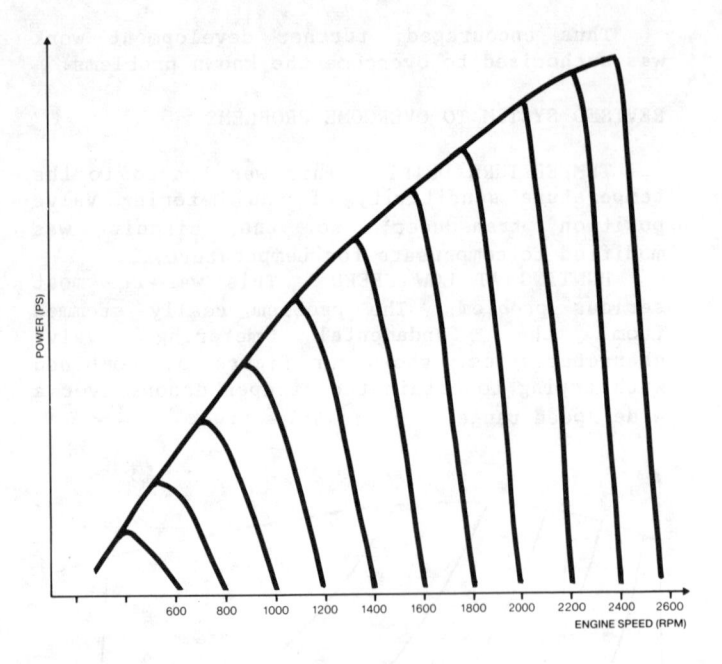

Fig. 4 - Governor Characteristics (Mechanically Governed DPA)

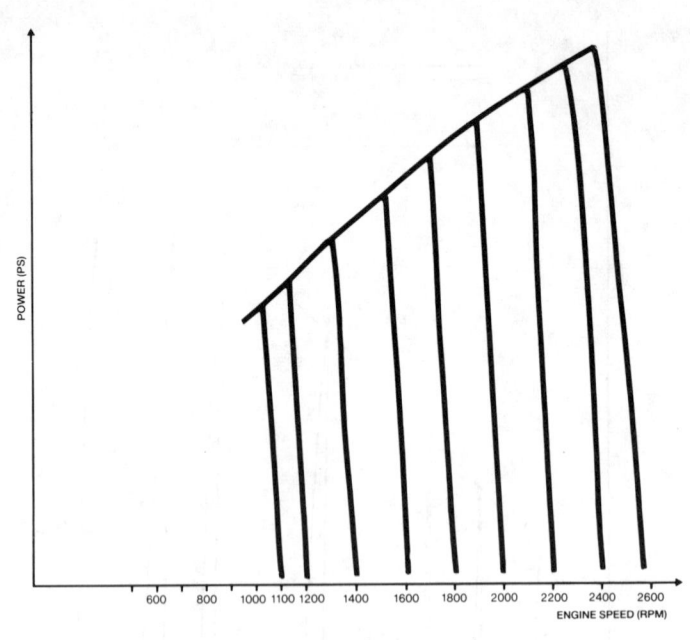

Fig. 6 - Governor Characteristics (Electronic Power Take Off Governor)

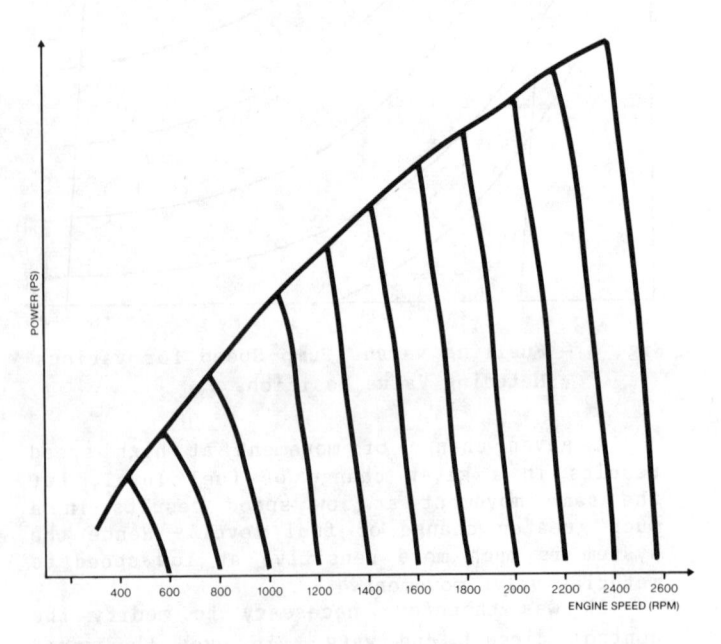

Fig. 5 - Governor Characteristics (Electronic Road Speed Governor)

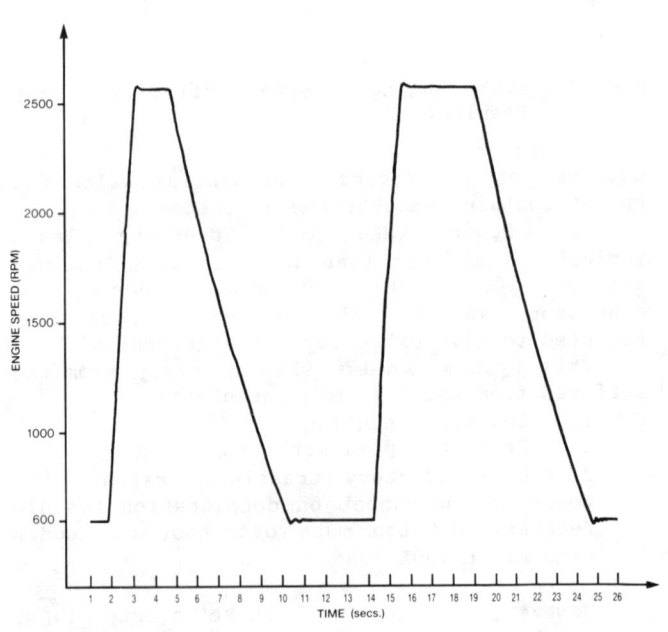

Fig. 7 - Free Engine Acceleration/Deceleration (Road Speed Governor)

TRANSIENT RESPONSE - For all of the governor pull off's and transient response work, a chart recorder was used which monitored both speed and metering valve position.

The engine was connected to an electric dynamometer, with arrangements made to have electric control of change of load, so that the load could be changed from zero to maximum and back at any speed. Additionally, a switch was incorporated in the system to try and simulate sudden load changes.

Figure 7 shows the acceleration and deceleration of the engine with the dynamometer disconnected, using the road speed governor. As to be expected, on this test there was no difference between this Road Speed and Power Take Off governor, both giving good overshoot and undershoot characteristics. The rapid settling time is noteworthy.

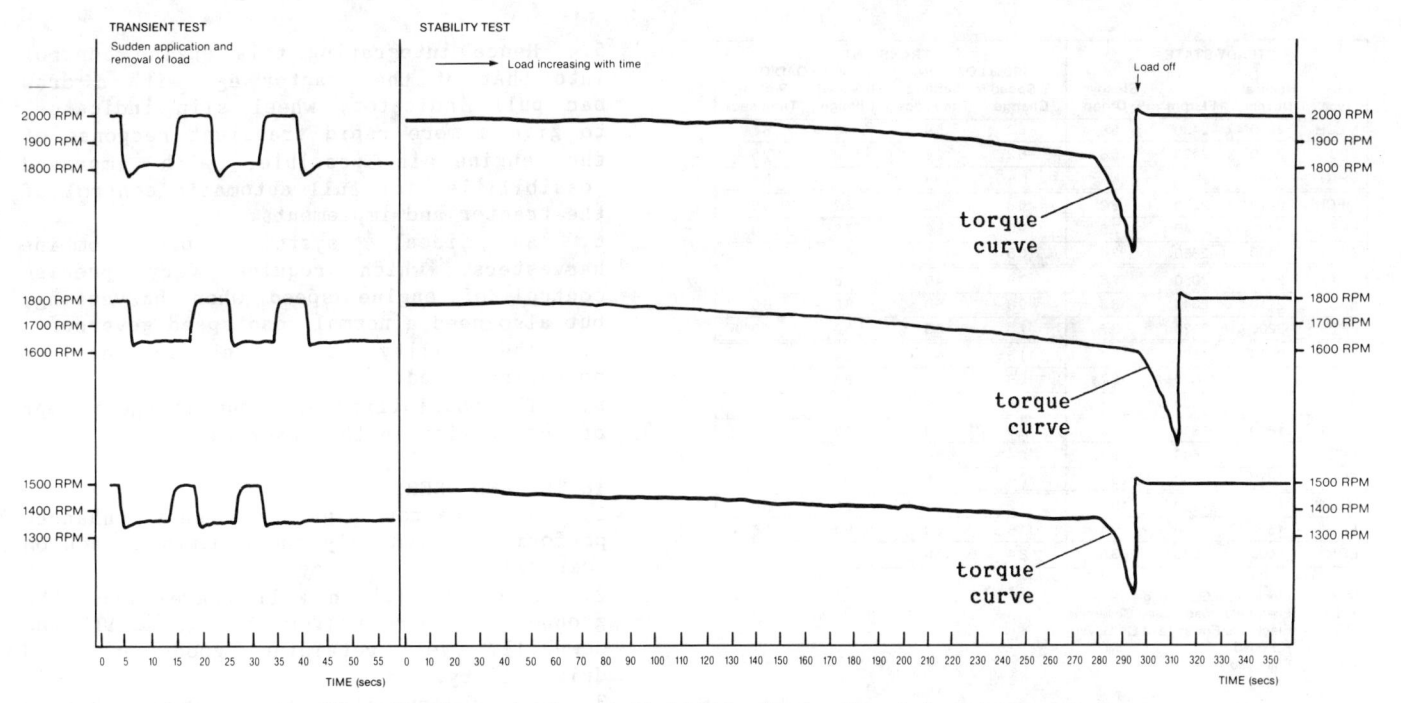

Fig. 8 - Transient and Stability Tests (Road Speed Governor)

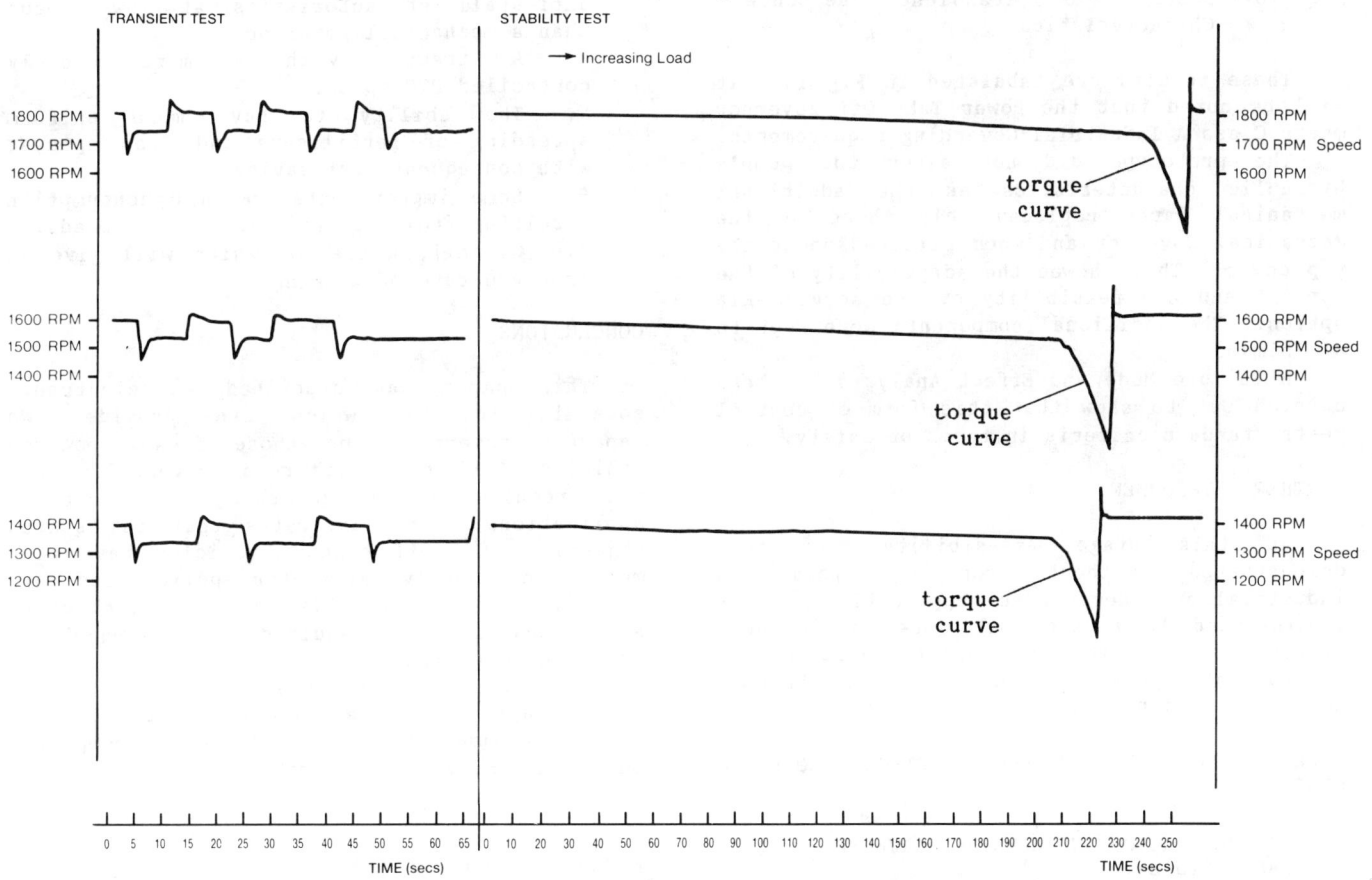

Fig. 9 - Transient and Stability Tests (Power Take Off Governor)

Shown in Fig 8 is the transient response with the road speed governor, and in Fig 9 is shown the transient response of the Power Take Off governor, for 100% load change.

(In this context 100% is assumed the load at which a generator would operate, which is approximately 85% of the normal full torque at that speed).

Gov. Type	STEADY STATE			TRANSIENT			
	Nominal NL rpm	FL rpm	Steady % Droop	100% LOAD ON		100% LOAD OFF	
				% Speed Change	Settling Time secs	% Speed Change	Settling Time secs
MECH	2570	2430	5.8	6.2	5.0	5.8	5.0
ERS	2560	2420	5.8	6.2	2.5	5.8	2.5
EPTO	NA	NA	NA	NA	NA	NA	NA
MECH	2410	2270	6.2	6.6	4.0	6.2	5.0
ERS	2410	2240	7.5	9.0	5.0	7.0	5.0
EPTO	2400	2320	3.4	6.8	2.5	4.3	2.5
MECH	2210	2050	7.8	7.8	4.0	7.8	5.0
ERS	2210	2040	8.3	9.3	5.0	8.8	6.0
EPTO	2200	2120	3.7	7.0	2.5	5.0	5.0
MECH	2010	1820	10.4	11.5	4.0	10.4	4.0
ERS	2000	1820	8.7	12.6	4.0	9.8	5.0
EPTO	2000	1930	3.6	6.8	3.0	4.6	3.0
MECH	1800	1610	11.8	13.0	5.0	11.8	3.0
ERS	1800	1640	9.8	10.4	5.0	9.8	2.5
EPTO	1810	1750	3.4	8.4	2.5	5.1	2.5
MECH	1610	1400	15.0	17.0	2.5	17.0	3.0
ERS	1500	1370	9.4	11.6	6.0	9.4	2.5
EPTO	1600	1540	3.8	8.4	2.5	5.1	2.5

Mech = Mechanical Governing
ERS = Electronic Road Speed Governing
EPTO = Electronic Power Take Off Governing
NL = No Load
FL = Full Load

Fig. 10 – Droop and Transient Response Characteristics

These results are tabulated in Fig 10. It will be noted that the Power Take Off governor meets Class A Industrial Governing requirements.

The prototype did not alter the pump's hydraulic characteristics as the additional mechanical parts replaced only those of the Mechanical Governor and were situated under the top cover. This showed the adaptability of the concept and the feasibility of the service kit option. The additional components were kept to a minimum.

A Failure Mode and Effect Analysis has been carried out to show that this form of control meets standard criteria in terms of safety.

FURTHER DEVELOPMENT

At this stage, feasibility has been demonstrated without applying this to industrial or generator set applications and it is intended to pursue such work in the near future. This governor concept could easily include a crystal reference for isochronous governing of generator sets.

EXPECTED BENEFITS FROM THIS ELECTRONIC GOVERNOR CONCEPT

TO THE TRACTOR AND COMBINE HARVESTER MANUFACTURER:
1. A simple form of electronic control which will enhance the performance of his tractor.
2. Simple to install.
3. The high pressure hydraulics remain unaltered.
4. Interface signals for speed and load can easily be provided if necessary, as also can diagnostic information.
5. Hence integrating this engine control into that of the tractor eg. with a draw bar pull indicator, wheel slip indicator, to give a more rapid transient response of the engine is possible, with improved possibilities for full automatic control of the tractor and implements.
6. An ideal system for combine harvesters, which require very precise control of engine speed when harvesting, but also need a normal road speed governor.
7. The ability to pre-set any desired operating speed.
8. The possibility of reducing the number of gear ratios in the tractor.

TO THE END USER:
1. A tractor having an enhanced performance with only the minimum of add-on complexity.
2. A tractor which will travel over the ground at a more uniform speed, uphill and downhill, yet retaining good on road driveability.
3. A governing system with better anti-stall characteristics at low speeds than a mechanical governor.
4. A tractor with a more closely controlled PTO speed.
5. The ability to have more uniform spreading of fertilisers and crop sprays, with consequent cost savings.
6. Some improvements in fuel consumption resulting from more uniform engine speeds.
7. Overall, a tractor which will give an improved rate of working.

CONCLUSIONS

This paper has described an electronic governing concept which can provide two separate governors for those diesel powered vehicles which need good road driveability but yet require close governing for certain applications, and illustrates that the stated objectives for off highway vehicles have been met. Additionally, generator applications have been shown to be possible, but more work using a generator is required to demonstrate isochronous governing.

The work has demonstrated the feasibility of such a system, as applied to Rotary DPA type injection pumps, which is the type of pump used on the majority of Tractors.

ACKNOWLEDGEMENT

The authors would like to thank the directors of Lucas CAV for permission to publish this paper, and colleagues within the Company who have given helpful advice.

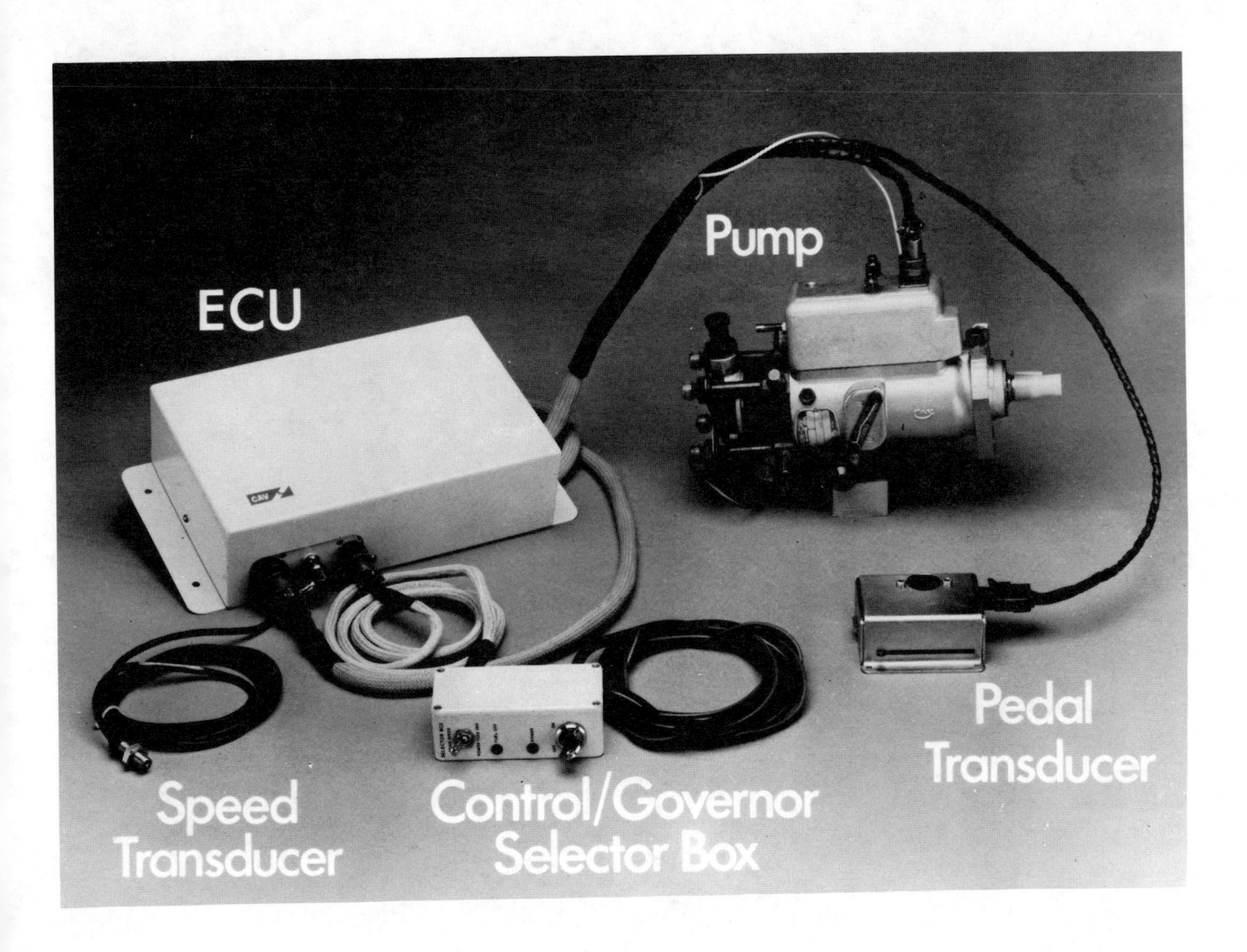

Fig. 11 - EGDPA System (Prototype)

REFERENCES

1. H Mayagi et al. "Toyota Electronic Control System for a Diesel Engine". SAE 830862.

2. M Wakabayashi, S Sakata, K Hamanaka, "Isuzu's New 12.0L Micro-Computer Controlled Turbocharged Diesel Engine". SAE 840510.

3. H Nakao, S Yamaguchi, "Electronically Controlled Engines for Passanger Cars". Journal No. 2 SAEJapan 1984/Vol 38.

4. M E Moncelle, G C Fortune, "Caterpillar 3406 PEEC (Programmable Electronic Engine Control)". SAE 850173.

5. P E Glikin, "An Electronic Fuel Injection System for Diesel Engines". SAE 850453